PUBLIC DATA NETWORKS

From Separate PDNs to the ISDN

Josef Pužman and Boris Kubín

PUBLIC DATA NETWORKS

From Separate PDNs to the ISDN

With 89 Figures

Springer-Verlag Berlin Heidelberg GmbH

Josef Pužman, PhD
PTT Research Institute, Hvoždanská 3, CS 149 50 Prague 4,
Czechoslovakia

Boris Kubín, PhD
PTT Research Institute, Hvoždanská 3, CS 149 50 Prague 4,
Czechoslovakia

ISBN 978-3-540-19580-1   ISBN 978-1-4471-1737-7 (eBook)
DOI 10.1007/978-1-4471-1737-7

British Library Cataloguing in Publication Data
Pužman, Josef, 1935–
  Public data networks: from separate PDNs to ISDN.
  I. Title   II. Kubín, Boris, 1923–
  004.6

Library of Congress Cataloging-in-Publication Data
Pužman, Josef.
  Public data networks: from separate PDNs to ISDN/Josef Pužman and Boris Kubín.
    p. cm.
  Includes bibliographical references and index.

  1. Computer networks.   2. Computer network architectures.   3. Integrated
  services digital networks.   I. Kubín, Boris.   II. Title.
TK5105.5.P9   1991   91-23893
384.3–dc20   CIP

© Josef Pužman and Boris Kubín 1992
Originally published by Springer-Verlag Berlin Heidelberg New York

Composition by Genesis Typesetting, Laser Quay, Rochester, Kent

34/3830-543210 Printed on acid-free paper

Preface

Computer techniques and microelectronics have permeated all areas of human activity. This has enormously influenced telecommunication systems: not only is the human being a direct user of telecommunication services but there is also a new kind of user – the computer and the terminal have provided the means for people to communicate with computers. Public networks support, far more conveniently than has been previously possible by means of private networks, efficient and less costly telecommunication between terminal and computer or between computers, regardless of who owns the data terminal on the serviced territory.

The pioneers of data teleprocessing utilized telegraph and telephone networks. However, it was not until the development of public data networks, CSPDNs as well as PSPDNs, that data communication became possible on a qualitatively higher level. The assortment of data services and user facilities gradually expanded, the quality of services improved, and new services appeared (datafax, teletex, MHS, EDIFACT, videotex). Network digitization and integration of networks and services have constituted a further qualitative change in the progress towards the ISDN. The ISDN uses advanced transmission and switching techniques with the aim of enhancing the telecommunication services provided to its users. An ISDN has much in common with the PDN as far as architecture, methods of network management and functions are concerned, but there is in addition a distinct change in the methods of access and signalling.

An exhaustive publication on networks providing data transmission services has not yet appeared. Perhaps the reason for this lies in the rapid development of the specialization which manifests itself in the fact that information can sometimes become obsolete before it can be published. Nevertheless, we believe in the usefulness of documenting the present technology, although we realize that the result of this work is only a snap-shot of the present state of the art. In the flood of implementation changes some principles remain unchanged: and there is even the rediscovery of old principles, such as store and forward message transfer, which was commonly used in early telegraph networks. The presentation of packet switched networks can also allow us to use the analogy of transporting mail in the postal network or of handling wagons in a railway network. Of course the reader is asked to realize that the novum in the application of old principles lies in the drastic compression of time scale (in the order of one million to one!) and in the liberation of information from such classical carrying media as paper.

Unlike most other books of this type, we have devoted a certain amount if space to the history of data communication which – though short – is interesting and yields very useful information. We would like the reader to get an idea of the development of data communication, and about the individuals and teams that participated in it.

There were certain difficulties with bibliographical references. Out of more than 1000 available sources (which form only a small fraction of what has been written about data networks and the ISDN) it was impossible to make a selection which would not give preference to only a few of them. Therefore we eventually decided to indicate the ISO standards and CCITT recommendations which were available by the beginning of 1991 (Appendices 3 and 4) and compile a selection of bibliographical references. First of all there are tutorials (books and special issues of some periodicals); there are several special papers for filling the gap caused by the necessary conciseness of this book; and, last but not least, there are already several classical works documenting the short history of telecommunication networks. We are aware that these references are not a representative sample but rather a heterogeneous fragment arranged only in alphabetical order, and for this reason references are not given in the text itself (with a few exceptions) but only under the headings of chapters, sections and subsections.

We believe that the knowledge and experience we have assembled can assist our readers in a better understanding of modern telecommunications, and give them the necessary support in making "future-proof" decisions.

June, 1991   Josef Pužman
                                                             Boris Kubín

Contents

Basic Abbreviations and Symbols Used in the Book

Although the authors admit that too many abbreviations might be a nuisance to the reader they feel that a preliminary knowledge of the most frequent ones (listed below) could be beneficial. The same applies to frequently-appearing symbols in the illustrations concerning data networks (see the figure). A comprehensive list of abbreviations is given in Appendix 5.

bit/s	Bits per second (unit of data signalling rate)
CSPDN	Circuit switched public data network
CCITT	International Telegraph and Telephone Consultative Committee
D.#	CCITT recommendation of the D series – General tariff principles – Charging and accounting in international telecommunication services (the full stop is followed by the recommendation number)
DCE	Data circuit terminating equipment
DTE	Data terminal equipment
G.#	CCITT recommendation of the G series – Characteristics of international circuits and transmission systems
I.#	CCITT recommendation of the I series – Integrated services digital network (ISDN)
IA5	International alphabet No. 5
ISDN	Integrated services digital network
ISO	International Organisation for Standardisation; standard approved by ISO (number of standard follows)
OSI	Open systems interconnection
PAD	Packet assembly and disassembly facility
PDN	Public data network
PSPDN	Packet switched public data network
PSTN	Public switched telephone network
Q.#	CCITT recommendation of the Q series – Telephone switching and signalling

R.#   CCITT recommendation of the R series – Telegraph transmission

RPOA   Recognized private operating agency

S   Segment (64 octets)

S.#   CCITT recommendation of the S series – Telegraph services terminal equipment

T.#   CCITT recommendation of the T series – Terminal equipment and protocols for telematic services

V.#   CCITT recommendation of the V series – Data communication over the telephone network

X.#   CCITT recommendation of the X series – Data communication networks

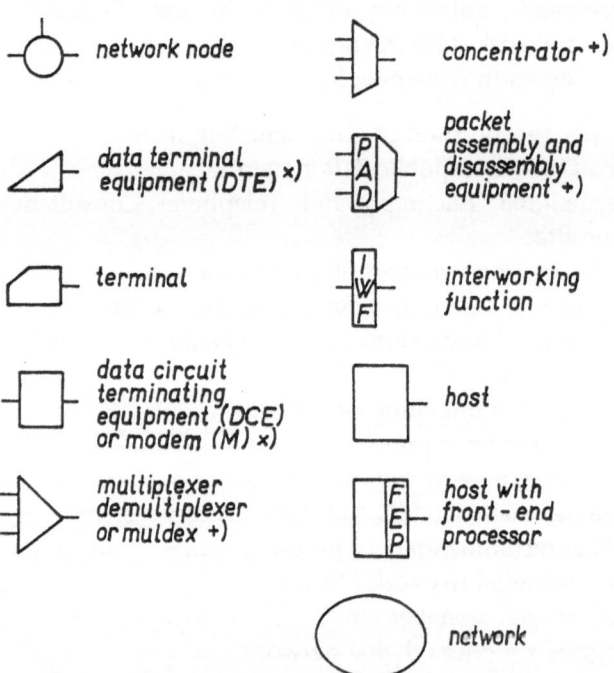

Symbols for schematic representation of public data networks and their constituent parts. x, if necessary, a letter indicating the type of equipment is added to the symbol; +, the symbol may be simplified by drawing only one inlet/outlet instead of three.

1 ■ Introduction

The development of telecommunications up to the middle of the present century was marked by the improvement of interpersonal telecommunication. In former days this applied to the electric transmission of written messages over a distance. In the earliest days of telecommunication messages were converted by coding into a form suitable for teletransmission. The principle of such conversion into an appropriate form (formalization principle) was much later extended so that it applied not only to information transmission but also to its processing and storage. In the 1950s this coded information came to be called "data".

The beginnings of the development of telephony towards the end of the last century (Fig. 1.1) led to the early recognition that it is economically intolerable and practically infeasible to interconnect telephone sets by individual lines. The solution to this problem was to connect all the telephones to a manual switchboard, or to a cluster of such switchboards (forming an exchange), in the centre of an inhabited area. The task of an exchange was to satisfy the demand for communication between two users by establishing a two-way transmission path (circuit) between them for the required time, responding to the initiative of one of them (the calling subscriber). This path is set up by switching, so that any pair of subscriber lines can be temporarily interconnected.

The interconnection of exchanges by groups of interexchange circuits created circuit switched telephone networks, covering larger and larger territorial areas. Public telephone networks in particular proliferated because they met the telecommunication needs of the area covered, using comparatively simple equipment.

Further enhancement of the telephone network, especially on the interurban and international levels, was stimulated by the automation of switching functions (Strowger's invention of the step-by-step switch in 1891) and the introduction of multichannel transmission systems based on frequency division. The ubiquity of the telephone and its adaptability as an interpersonal communication tool continues to affect the development of other branches of telecommunication, including data communication.

Telegraphy was forced into the background by the upsurge of the telephone network so that towards the end of the nineteenth century it was virtually limited to the public telegraph service for the transmission of telegrams. It was kept alive only because of several features different from those of telephony. These were that it does not need the presence of the

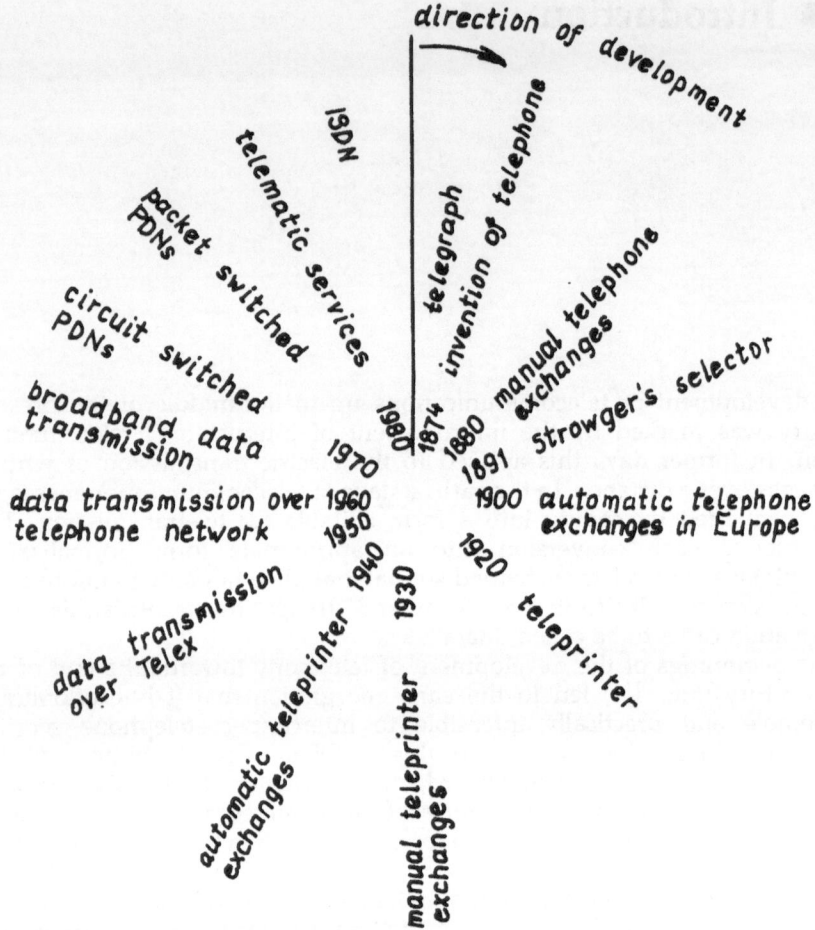

Fig. 1.1. Evolution of telecommunication techniques and data transmission. PDN, public data network; ISDN, integrated services digital network.

addressee during reception and it easily overcomes language barriers and difficulties arising from time-zone differences. Even the public telegraph tried to make use of the transmission capacities of the network by concentrating its functions into exchanges (telegraph offices), but switching was performed by the reception of telegrams and their sending (manual and later automated retransmission) to the required destination. In this case circuit switching is substituted by a new function – telegraph message switching depending on the addresses incorporated into the heading of the messages.

For the effective growth of alphabetic telegraphy it was necessary to make the telegraph apparatus available to the user on his site and to simplify the operator's work. That was made possible at the beginning of the twentieth century by the invention of the teleprinter with a keyboard and a character printer as its main parts – it resembled an office typewriter.

The 1920s brought the concept of the teleprinter network with switch-boards and exchanges functioning in the same way as exchanges in the telephone network. The main method of distant (interurban) transmission of teleprinter signals was by using the multiplexing on the telephone channel by frequency division. Since 1976 this has been replaced by time division. The global teleprinter exchange network – Telex – has culminated in having over one-and-a-half million subscribers and was characterized early on by a high degree of automation of its exchanges. The signalling rate of its telegraph channels (50 bits/s) is matched to the speed of typing. The familiar storage medium has been the five-track punch tape. This has been gradually replaced by the floppy disk which is also used at the sending side, where it speeds up transmission and thus economizes on charges.

Another area of interest when analysing the conditions which led to the origination of public data networks (PDNs) is the evolution of data teleprocessing (Fig. 1.2). This is closely related to the development of computer technology.

The growing demand for more and better computing techniques after the Second World War led to the concentration of computing activities into computing centres. As a rule, the input information for processing by these centres was originated in remote places. Therefore it was recorded in the place of origination (on the basis of direct readings from measuring apparatus or primary documents) in a manner suitable for subsequent processing, thus

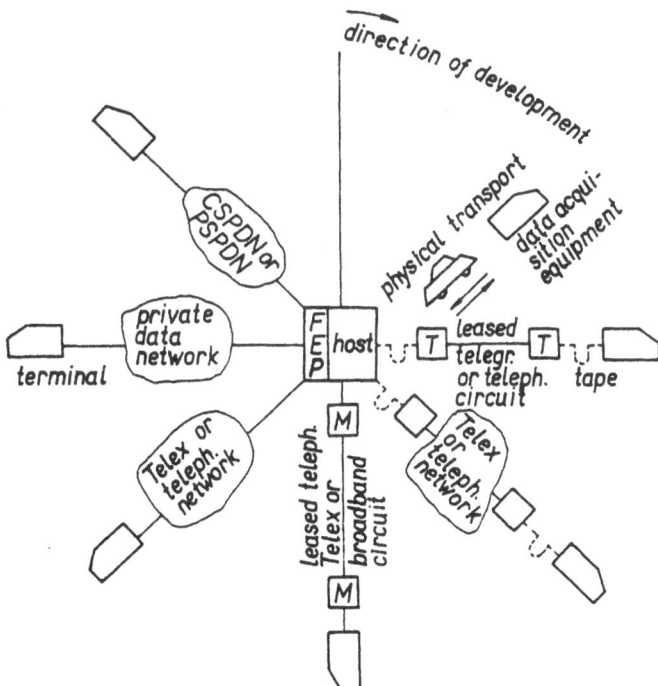

Fig. 1.2. Evolution of data teleprocessing. FEP, front-end processor; T, teleprinter; M, modem; CSPDN, circuit switched public data network; PSPDN, packet switched public data network.

producing the required input data. The punch card and the punch tape were commonly used as the recording media for data acquisition. These media had to be transported to the computing centre. For the bulk processing of data, and when the prompt availability of processing results was not important, the transport of the media containing the recorded data by messenger or by post was sufficient. The results of processing were transferred, usually from the computing centre to the place where the data had to be applied, in a similar manner.

To cut down the response time between data origination and the production of the results of data processing, in the early 1950s data teleprocessing over teleprinter networks was introduced. This presented no difficulties and no significant investment was needed to make a telecopy of a punch tape, even though a throughput of only 24 000 characters per hour was achieved and there was no protection against transmission errors.

In the 1960s the public switched telephone network (PSTN) developed a similar procedure. The telephone subscriber was equipped with a modem enabling the transmission of data signals over the telephone network by voice frequencies compatible with the standard telephone channel (300 Hz to 3400 Hz). On the sending side the modem was fed by an eight-track tape reader and on the receiving side it produced signals for the input of an eight-track tape punch (reperforator). The transmission took place in frames and was protected by a double parity detection code. In the event of error detection the transmission was repeated. The establishment of connections in the telephone network was made by the conventional method using the telephone apparatus. The modem had two functions: as signal converter, for data signal transposition to the telephone band and vice versa, and to connect the telephone line either to the telephone apparatus or to the data terminal equipment (DTE; the common name for computer and terminal connected to a telecommunication network). It goes without saying that a service similar to that provided by switched telegraph or telephone connections can be achieved by the leased telegraph or leased telephone circuit. Compared with the switched connection it has the advantage of permanent availability and better transmission quality because it is possible to assemble it from sections having known and adjustable characteristics. Therefore such a circuit supports higher data signalling rates.

On leased telephone circuits data signalling rates up to 14 400 bit/s are used at present, compared to rates of up to 9600 bit/s on a switched connection. However, the use of a permanent circuit is economically justified only in cases where the usage is high. Hence the endeavour to organize transmission in such a way as to exploit the leased circuit as much as possible: by connecting to it several terminals and controlling communication by a front-end processor (FEP). The situation when the customer himself controls communication on a leased circuit, because of the inadequacy of switched connections for his needs, signals to the service provider that he should handle this problem within his strategy of planning the data communication services.

Let us resume the review of the development of data teleprocessing. The off-line method of transmission whereby data was temporarily recorded on a medium was – for a number of teleprocessing applications – rather slow and inconvenient due to the need to use punch cards or punch tape in large quantities. This problem affected the off-line process on the data acquisition side as well as the transfer of data from the telecommunication circuit to the computer input and vice versa on the opposite side. The delay of off-line operation proved to be unacceptable in cases where the cycle "data origination – data processing – application of results" had to be run in real time or when the subscriber wished to communicate by means of a terminal with the distant computer in conversational operating mode.

Another feature introduced in the 1960s was the on-line operation between the leased circuit or switched connection on one side and the computer or terminal on the other side. The response time in data teleprocessing was cut down drastically. The direct (on-line) connection of the computer and its corresponding terminals to the switched telecommunication network enabled the terminal operator at any time to contact the computer almost immediately (in the case of full network automation) with a request for computation, information retrieval or fulfilment of other tasks. Companies with territorially dispersed enterprises and administrative headquarters equipped with a computing centre thus needed to obtain private terminal networks.

At first sight it would seem that a terminal network on the basis of leased or switched circuits of the telegraph type or telephone type is an ideal solution to all teleprocessing problems. However, this assumption has proved to be wrong. The rapid development of computer technology imposes more and more exacting requirements on both the quality and quantity of data teleprocessing systems. In the course of time computers have become more and more sophisticated and have increased dramatically in number within various branches of both the national economy and human activity as a whole. They penetrate down to our workplaces, into households and even into private life.

The original assumption of the overall advantage of the concentration of processing capabilities and memory resources into large computers and of their utilization by dispersed users by means of telecommunication networks failed. The merit of equipping smaller organizations with computers and of entrusting them with the bulk of local data processing became evident. Only jobs with which the local computer cannot cope require the assistance of a master computer or another computer of its own level, to make use of its processing and/or memory resources and possibly also to borrow its software and access its data files.

The intercomputer communication needed to achieve such co-operation is again provided by telecommunication resources made available at the required time, in satisfactory quality and quantity. In this way computer networks combined with terminal networks are created. Even in such an array parts of the tasks are dealt with by the terminals themselves if they have sufficiently "intelligent" software (hence the term "programmable terminals").

In this environment the means of telecommunication have great significance but exhibit some shortcomings caused by the very fact that they had been conceived only for interpersonal communication. In comparison with the internal data signalling rates between the constituent parts of a computer (of the order of 10^{10} bit/s), conventional telecommunication means are relatively slow (of the order of 10^4 bit/s). There is a similar discrepancy between the mean duration of a conventional call in the telephone or Telex network (several minutes) and the duration of one data call (for example, a data message of 10 000 bits will be transmitted via a digitized telephone channel at 2400 bit/s in about 4 s). In addition, the short holding time of a switched circuit engaged in data transmission contrasts with the call establishment time in conventional telecommunication networks (several seconds or tens of seconds).

The inadequacies of using conventional telecommunication networks, described above, for data teleprocessing resulted in the abandonment of the use of Telex circuits for data transmission and the search for ways of providing faster data channels. An interim solution involved the digitization of broadband channels of multichannel carrier telephony by broadband modems. An example is the 48 kbit/s or 64 kbit/s channel in the frequency band of the primary group designated for the allocation of 12 standardized telephone channels by frequency division. The best way to obtain adequate

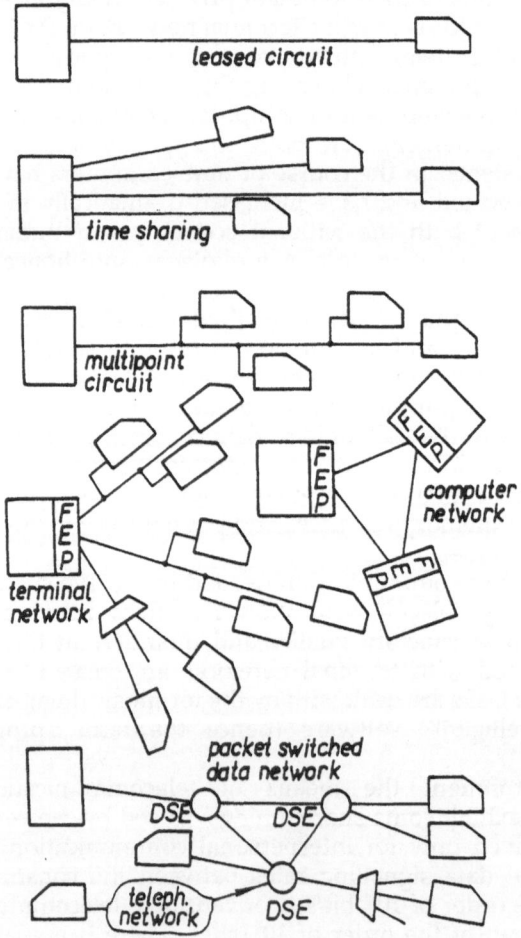

Fig. 1.3. Evolution leading to the packet switched public data network. FEP, front-end processor; DSE, data switching exchange.

fast digital channels for data transmission and other digital applications is the use of a channel with an ample frequency band (in coaxial cables, microwave links or optical fibre), the digitization of this channel and the acquisition of data channels with the required speeds by subsequent time division.

The development of the terminal-to-computer communication technique (Figs 1.3 and 1.4) has gone through several stages but two basic approaches to the issue can be identified: circuit switching and packet switching. The former resulted in the development of circuit switched public data networks (CSPDNs), the latter in packet switched public data networks (PSPDNs).

The first steps towards CSPDNs can be seen in the perfecting of teleprinter exchanges after 1960. Electronic control and the sophisticated technology of the switching system in the switching network of a Telex exchange brought about the ability to increase the modulation rate of telegraph type exchanges from 50 baud to 200 baud and to admit a more comprehensive code than the five-bit code standardized for Telex.

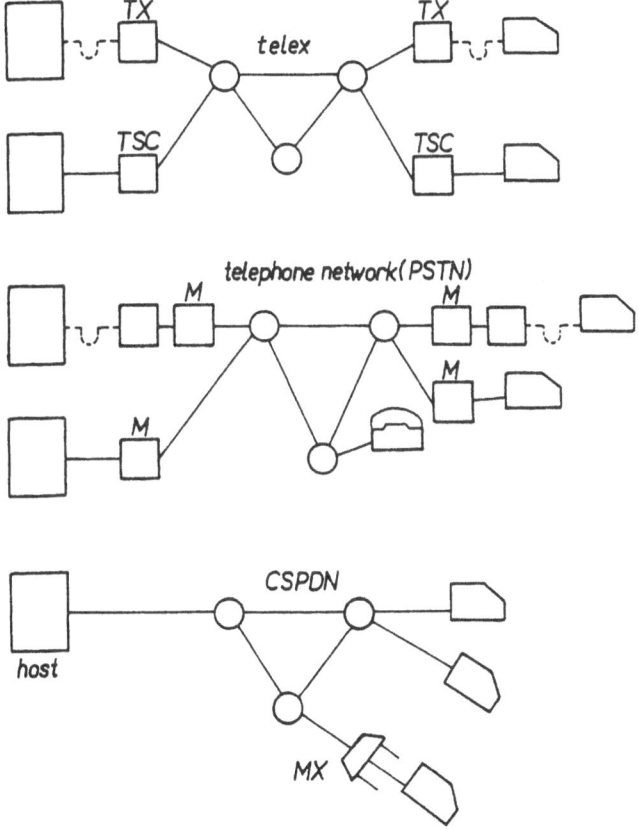

Fig. 1.4. Evolution leading to the circuit switched public data network. TX, teleprinter; TSC, telegraph signal converter; PSTN, public switched telephone network; M, modem; MX, multiplexer.

Even more favourable conditions were created in the 1970s when microelectronics and electronic control created the conditions for routing digital signals across the network without having to use any mechanical switching elements. The exchange assumed the form of a computer with many inputs and outputs. Its task is reduced to transporting bits or groups of bits from an input to an output designated by a stored address. It behaves as if it interconnected the corresponding telecommunication circuits by mechanical switches. In contrast to classical exchanges, it is capable of regenerating the signal transmitted through the switching field.

The introduction of CSPDNs began on the national level, often in combined text and data transmission exchanges within the so-called integrated data networks. Interworking between CSPDNs on the international level for CSPDNs supporting start–stop DTEs was standardized in 1972 and for CSPDNs supporting synchronous DTEs in 1976. The creation of a global CSPDN system had to be preceded by the unification of national systems in DTE–network interface standards for both start–stop and synchronous DTEs in 1972.

The packet switching approach was based upon the analysis of the needs of computer-to-computer and terminal-to-computer telecommunication and the evolution culminated in the global PSPDN. The first development in on-line data teleprocessing (distant processing without the use of punch tape) was a single terminal connected to a distant computer by a leased circuit.

The next development stage was the terminal network, where the computer communicated with a number of distant terminals simultaneously on a time-sharing basis. This occupied the processing capacity of the computer and the usage of the connecting lines was relatively poor.

The introduction of a multipoint circuit results in high usage of the line but does not relieve the computer of the control procedure. This drawback is considerable if several multipoint circuits are connected to the computer. Therefore the FEP is responsible for the actual telecommunication control but the main computer "lends" its processing capacity to the terminals. In other words, the "intelligence" of the terminals resides in the main computer, hence its name "host computer".

To cover a large area several computers are linked together by leased circuits to form a combined computer and terminal network. The network control functions are taken care of by FEPs.

The experience gained in telegraph networks with message switching (where the store-and-forward principle was applied to the routing of telegrams) led to the idea of applying this principle to data transmission. However, the message switching system had to be modified according to the requirements of efficient data communication. This modification consisted of splitting long data messages into shorter fragments of limited length and supplying them with adequate accompanying information to ensure their optimum passage through the network. An analogy with the organization of the movement of mail items in the postal network was developed at an early stage. The term "packet" was given to the duly equipped fragment. Any information including commands and responses for communication control had to be fitted into packets. Part of the communication control was transferred into nodal computers at switching nodes which, together with data circuits, formed a communication subsystem for host computers, FEPs and terminals.

   In the second half of the 1970s computer manufacturers designed standard architectures for the formation of application-independent packet switched private data networks (user data networks). The philosophy of such a network is based upon a consequential separation of the data processing functions and the data transport functions. This created favourable conditions for designing data networks, open to any DTEs that fulfil the connection requirements assessed by CCITT (Comité Consultatif International des Télégraphes et Téléphones) recommendations and ISO (International Organisation for Standardisation) standards (hence the term "public data networks"). They are able to incorporate the private networks described above. 1978 can be regarded as the year of birth of the global PSPDN, because it was then that the CCITT issued recommendation X.25 for the interface between the packet-mode DTEs and the PSPDN, and X.75 for terminal and transit control procedures and data transfer systems on international circuits between national PSPDNs.

   A relevant prerequisite for the build-up of PDNs is the existence of an expanded digital transmission network from which high quality and cheap digital channels can be made available to the PDN designer. This transmission network on all international and national levels of the telecommunication network is based on different transmission media, of which the most "future-proof" are optical fibres and satellite links.

   Beside its main function of providing transmission for data communication, the PDN as a purely digital network can host telematic services (teletex, telefax, FAX 4, bureaufax and videotex) in the capacity of a host network. These services, defined as new telecommunication services excluding telephone, telegraph and data transmission services for the purpose of exchange of information via telecommunication networks, were officially introduced on the international level by CCITT Resolution No. 13 in 1980.

   The advent of the integrated services digital network (ISDN) brought about the requirement to include data transmission services among the services provided by means of this network. Because of the digital nature of the ISDN such an integration is possible without any problems for circuit switched user classes. To be capable of achieving packet switching, an ISDN exchange must be equipped with a special packet handling facility, the packet handler (PH).

   A specific feature of the evolution of telecommunications for data processing is the survival of its historical phases up to the present time, a phenomenon explained by their economic and functional merits. This statement is valid for the physical transport of data in data teleprocessing systems as well as for the use of leased and switched circuits of teleprinter networks and telephone networks, both public and private. In spite of the fact that the PDNs (as dedicated networks or networks integrated into ISDNs) are the leaders in this evolution, they can promote themselves only if they are economically attractive or otherwise advantageous for the network user as well as for the service provider. An example of the persistence of conventional data transmission methods is the widespread use of modem-equipped leased telephone circuits for data transmission, in preference to the commissioning of PDNs.

   An important issue being raised by telecommunication futurologists is the *raison d'être* of PDNs in view of the ascension of ISDNs, because ISDNs include – by definition – all telecommunication services including data transmission and all the other services hitherto hosted by PDNs and provided on a "one pair of wires" basis. It is probable that in a pure ISDN environment, PDNs will not vanish but will reappear integrated into the ISDN system. In an interim period the ISDN can play the role of an access network for attracting distant PDN users. This function will survive even in the later stages of ISDN development, especially as to the PSPDN functions, because the PH facilities will reside only in exchanges of a higher network level.

Observation of historical evolution and knowledge of the present state of development of PDNs leads us to assume that these networks will continue to develop in quantity and quality in coexistence with, or within, the ISDN.

2 ■ Public Data Network Principles [22,35,70]

2.1 Basic Terms

The main function of a PDN is to enable communication for computers: computer-to-computer telecommunication and terminal-to-computer telecommunication. The general term *communication* stands for a process in which two or more parties are involved with the purpose of raising the level of knowledge of at least one of them. *Information* is considered to be the difference in knowledge before and after communication. In communication we can always identify the originator of the information represented by the information source, the direction of information transfer, and the receiver of the information represented by the information sink.

Telecommunication is communication over a distance requiring the use of the propagation of electromagnetic waves in free space, in waveguides (including optical fibres) or along conductors. The physical quantity involved in the transfer of information from source to sink is called a *signal*.

The signal is generated in the signal transmitter and received by the signal receiver (Fig. 2.1). They terminate the telecommunication or transmission *channel*, which is defined as the means of signal transmission between two points in one direction for the transfer of information, the source and sink not necessarily coinciding with these two points. The definition of the term *circuit* (short for "telecommunication circuit") differs from the definition of the term *channel* only in that it involves the transmission of signals in both directions and implies that the information source and information sink on either side of the circuit are related by some sort of association, generally by being situated in the same place or station and by belonging to a single communication session.

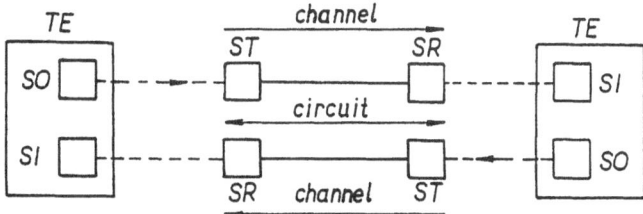

Fig. 2.1. Basic telecommunication terms. SI, information sink; SO, information source; SR, signal receiver; ST, signal transmitter; TE, terminal equipment.

A computer in the context of this book stands for any task-oriented information processing equipment. A terminal is a piece of equipment enabling the user to communicate with a computer. For a computer and a terminal connected to a network we use the common term *data terminal equipment* (DTE).

There is a general consensus to use the term *data* for information in a form suitable for processing, storage or transmission. Data as an adjective denotes a quality, activity or capability related to data. Thus *data transmission* means transmission of data; *data communication* is communication based upon data transmission; a *data network* is a network primarily used for data communication; a data channel is a channel for data transmission; a data circuit is a circuit for two-way data transmission; a data source is an information source generating data.

By offering its services to all users asking for the service provision in a certain area, for example the territory of a country, a network assumes the attribute "public". The opposite of a public network is a private network which provides services for a limited group of users. Private networks can be hosted by and connected to public networks.

In most cases modern data communication consists of the transfer of data read from the memory of one DTE to the memory of another DTE. The term "reading" implies that data in the data source is not necessarily being deleted during the transfer process. The data source and the distant data sink are linked by a data link whose constituent parts are defined in Fig. 2.2, where the terms in Fig. 2.1 are modified to apply to data communication.

Because of the great variety of DTEs the provision of a complete data communication service (belonging to the so-called teleservices, described later) would be a very difficult commitment for the public telecommunication service provider (a telecommunication administration or a recognized private operating agency, RPOA). Therefore the provider's basic task is to support data communication by providing the so-called transmission service or bearer service consisting of data transmission between DTEs connected to the

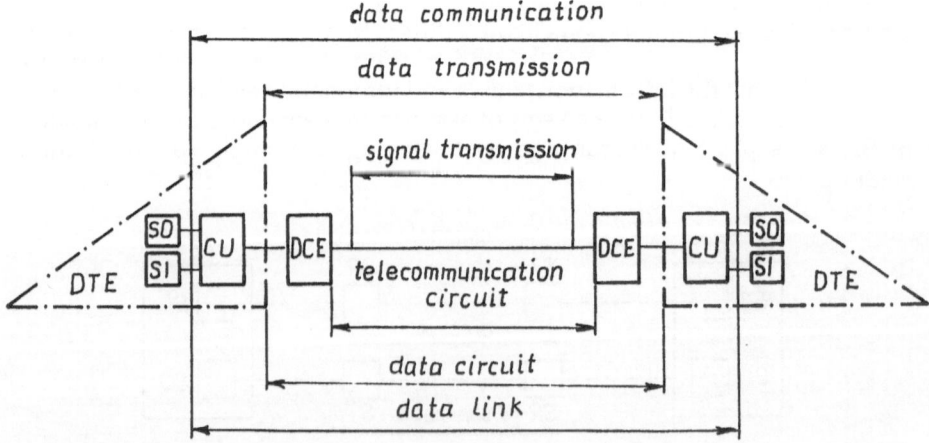

Fig. 2.2. Basic data communication terms. CU, communication unit; DCE, data circuit terminating equipment; DTE, data terminal equipment; SI, information sink; SO, information source.

telecommunication network. To this aim, the interface between the telecommunication network and any DTE must be adequately specified.

By *interface* we usually mean a boundary shared by two associated systems or equipments. When describing the relation between the PDN and its user, the equipment consists of the DTE and the corresponding equipment in the telecommunication network. The latter equipment is usually called *data circuit terminating equipment* (DCE), a term based upon the previously explained general telecommunications terminology.

The application of the term circuit in data communication leads to the term *data circuit* with two equivalent definitions:

- A pair of channels with opposite directions of data signal transmission between two points for associated data source and data sink on either side of the channel pair
- The means of data signal transmission in both directions between two points for associated data source and data sink on either side of the pair of points

The basic function of a PDN is to support data communication by data transmission between two DTEs connected to this network and requesting the communication. For this purpose the PDN may establish a data circuit between the corresponding DCEs. However, the definition of such a circuit is very loose because data transmission can be restricted to the transport of formalized messages between the DTEs. In this case we have to abandon the idea of a data circuit placed permanently at the disposal of the communicating DTEs for the communication period.

For ease of explanation it is convenient to consider the PDN as a "black box" presenting its DCEs to the users' DTEs all over the area provided by the data transmission service. This general model implies two stages of communication (Fig. 2.3): communication between the DTE and its corresponding DCE in the PDN and communication between two DTEs connected to the PDN, which is the ultimate purpose of any PDN.

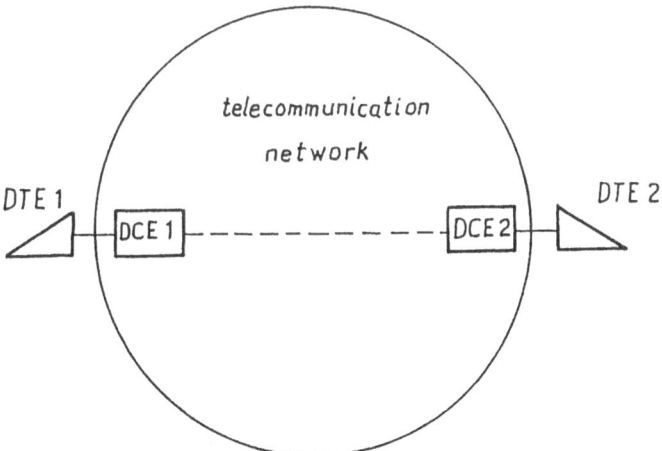

Fig. 2.3. The principle of data communication and data transmission over a telecommunication network. DTE, data terminal equipment; DCE, data circuit terminating equipment.

The black box concept of a telecommunication network can be based on the fact that it is a set of means (hardware as well as software, transmission as well as switching) supporting the provision of telecommunication services. The internal organization of any telecommunication network defined by its structure and rules of operation is designed with the aim of attaining economy by sharing transmission resources. This broader definition of a telecommunication network applied to a data communication network involves the system of permanently leased data circuits as well as data circuits realized on a telephone network by the use of modems. Beside PDNs there are private data networks utilizing either private telecommunication circuits or leased circuits obtained by utilizing the leased circuit service provided by the telecommunication administration or RPOA.

The path of evolution leads to the provision of PDN services by the integrated services digital network (ISDN). This is evident from the very name of this network with the assumption that it supports the provision of data transmission services in the same way as a PDN. An important feature of a PDN is the transmission of digital signals. A digital signal is a signal encoded as elements corresponding to digits of a number system. In other words, the digital signal is a sequence of signal elements. A signal element is an elementary part of a signal distinguished from the others by one or more characteristics such as its nature, magnitude, duration and relative position.

Data transmission for data communication is not the only telecommunication service supported by PDNs or equivalents. The ability to transmit digital signals predetermines a PDN to support those telecommunication services which, like data communication, are based upon the transmission of information between the memories of terminal equipment but include the provision or at least specification of the terminal equipment as well. These are the so-called teleservices: complete telecommunication services (as opposed to bearer services such as data transmission) and the informatic feature (exchange of information) has produced the more specific name *telematic services*. They include not only the modern text communication service (teletex), modern picture telecommunication services (facsimile services) and the terminal-to-database telecommunication service (interactive videotex), but also other services not yet standardized. Another important role of a PDN or its equivalent is the support of the message handling system (MHS), based upon the use of computers and terminals distributed over a given area. The system enables subscribers to exchange messages on a store-and-forward basis. The two main services provided by the MHS are interpersonal messaging (IPM) and message transfer (MT), which supports general, application-independent, message exchange. The hosting of teleservices in addition to the basic data transmission service is a welcome and important source of revenue for the PDN provider. This source sometimes yields a profit exceeding that from the data transmission services. These considerations justify the inclusion of PDNs in the category of value added networks (VANs).

A local area network (LAN) whose primary role is to enable communication between a specific user's DTEs at his premises can be connected to a PDN at its DTE, thus giving the LAN access to all other DTEs (including LANs elsewhere) and contributing to the establishment of wide area networks (WANs) and metropolitan area networks (MANs).

A relevant aspect of data communication is the incorporation of a PDN or its equivalent into the overall telecommunication and local communication environment. In addition, other public telecommunication networks such as PSTN, Telex, gentex, a different type of PDN (for example, packet switched as opposed to circuit switched) as well as the ISDN can have DTEs connected to them. This calls for the establishment of gateways and interworking units (IWUs) which enable DTEs to access the PDN (Fig. 2.4). The same philosophy applies to the relation between the PDN on the one hand and private networks, private branch exchanges and LANs generally limited to one user or user group on the other hand. The need for communication between DTEs connected to PDNs in different countries creates the necessity of interworking between PDNs – even of the same type – on the international level. This interworking is a typical application area of international standardization. A relatively new feature of CCITT recommendations in this area is that they also deal with the interface between a network and a DTE. It is not a national matter (as it might seem to be at first sight) because international data transmission involves the transmission of data signals between DTEs in different countries.

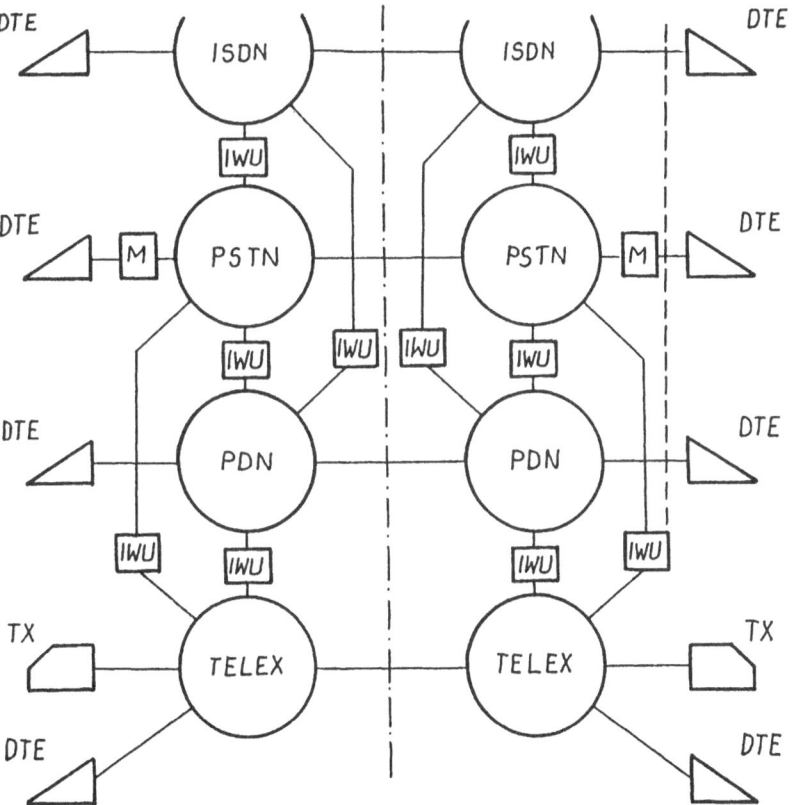

Fig. 2.4. Interworking between telecommunication networks supporting DTEs. IWU, interworking unit; M, modem; TX, teleprinter (operated as a terminal).

A telecommunication network dedicated to the provision of certain telecommunication services (telephony, Telex, data transmission) can be called a dedicated telecommunication network. Progress in digitization of transmission and switching has brought about the possibility of building a digital network for the provision of all telecommunication services hitherto provided by dedicated networks. The term for this new kind of network – *integrated services digital network* – is self-explanatory. The basic terms defined in this section assume point-to-point communication. The definitions could be extended to multipoint and point-to-multipoint if several DTEs are involved in a single communication session.

2.2 Data Transmission in PDNs

A PDN is a telecommunication network specialized primarily for data transmission between DTEs and operated by a telecommunication administration or RPOA. It supports data communication between various DTEs connected to the network directly or via other telecommunication networks, such as a PSTN, the Telex network or a PDN of a different type (for example, access to a PSPDN via a CSPDN).

PDN principles are based upon the specification of data transmission requirements for communication between DTEs via a PDN as given by the CCITT X series recommendations. This specification includes the rules of access of DTEs to the PDN: direct access via permanent circuits or access through other networks as summarized in Fig. 2.5. Different data transmission services are known according to the two types of PDNs (circuit switched, packet switched) and the ISDN. This also affects the classification of DTEs into user classes of service. Further classification criteria are data signalling rate (bit rate) and mode of operation (start– stop, synchronous) as shown in Table 2.1, with a survey of user classes of service for PDNs.

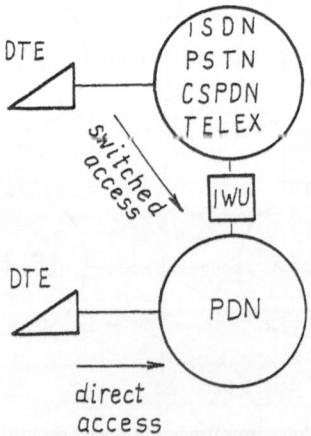

Fig. 2.5. Modes of access of DTEs to a PDN. IWU, interworking unit.

Table 2.1. Data transmission services (DTSs) and user classes of service in PDNs and ISDNs

DTS	Type of DTE	DTE–DCE interface (CCITT recommendation)	User class of service	Data signalling rate (bit/s)	Character structure in data transfer phase (a)[b]	Call control signals in the call control phase — Character structure (a)	Code
Circuit switched and leased circuit DTS	Start–stop mode	X.20 or X.20bis	1	300	11	11	IA5
			2	50[a]	7.5	11	IA5
				100[a]	7.5		IA5
				110[a]	11		IA5
				134.5[a]	9		IA5
				200	11		IA5
	Synchronous mode	X.21 or X.21bis	3	600			IA5
			4	2 400			IA5
			5	4 800			IA5
			6	9 600			IA5
			7	48 000			IA5
			19	64 000			IA5
Packet switched DTS	Synchronous mode	X.25 or X.32	8	2 400			
			9	4 800			
			10	9 600			
			11	48 000			
			12	1 200			
			13	64 000			
	Start–stop mode	X.28	20	50–300	10/11		
			21	75/1 200	10		
			22	1 200	10		
			23	2 400	10		
DTS in ISDNs	Synchronous mode	I.411	30	64 000			

[a] In the call control phase the data signalling rate is 200 bit/s.
[b] a = unit interval of the start–stop signal.

The specification of the DTE–DCE interface has resulted in an important and comprehensive set of requirements for PDN functions and their reactions to environmental stimuli. Generally speaking, the interface specifications comprise all communication layers, though data transmission itself (not data communication or other teleservices based upon data transmission) engages only three of the seven layers of the Open Systems Interconnection reference model (OSI/RM) – the physical layer, the link layer and the network layer.

The DTE–DCE interface is internationally standardized – as distinct from the interface between the telephone and the PSTN – because this standardization enables interworking between DTEs connected to PDNs in different countries.

There are two principal ways of achieving data transmission in a PDN: *circuit switching* and *packet switching*. In the case of circuit switching, prior to data transmission a circuit is established for data transmission between the DTEs asking for mutual communication. This so-called complete data circuit is composed of two connecting circuits (subscriber circuits) and one or more interexchange circuits (trunk circuits) (Fig.2.6). The establishment of a complete data circuit involves the interconnection of the circuits in exchanges. For the purpose of this preparatory process the exchanges have to communicate with each other, and the terminal exchanges (as opposed to transit exchanges) communicate with the subscriber stations. Once a through-connection is established the communication is handed over to the two communicating DTEs. The procedure is similar to that in a PSTN or – even more so – in the Telex network. However, in a CSPDN, as distinct from the PSTN or Telex network, the call establishment durations are much shorter (tens of milliseconds) and the transmission quality is enhanced. If error correction is necessary, it is carried out by the DTEs. The operating mode and bit rate of the communicating DTEs must be the same as, and equal to, those of the data channels but not necessarily the same in both directions of transmission (for example, user class 21).

The operating principles of a packet switched PDN have been chosen in an attempt to remove some of the basic disadvantages of CSPDNs. Thus a PSPDN allows interworking between DTEs of different bit rates, optimally utilizes the network's transmission media, performs error correction of data

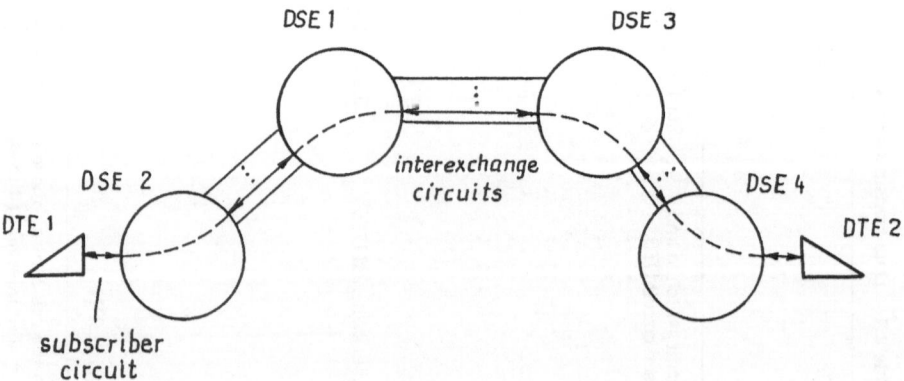

Fig. 2.6. The principle of a circuit switched PDN. DSE, data switching exchange.

during their passage through the network, avoids traffic congestion by data flow control and possesses a high degree of resilience in the event of network element failures.

These remarkable features are achieved by:

- Assembling data into packets (this operation resulting as a rule from disassembling data messages into fragments)
- Applying the store-and-forward principle together with a convenient routing strategy to the movement of the packets through memories in the data switching exchanges of the network (Fig. 2.7)
- Providing the packets with the address of the destination DTE and/or with data necessary for subsequent routing
- Subjecting the packets to error correction and checking for loss after their travel from memory to memory
- Limiting the flow of packets into the network from DTEs and possibly also between packet switching exchanges, to avoid traffic congestion
- Subjecting the packets that have reached the destination DTE to packet disassembly – an operation inverse to that of packet assembly

It is to be noted that if the DTEs are not capable of packet assembly and disassembly this task is delegated to a special facility – packet assemblers/disassemblers (PADs) – in the network. This applies to those types of start–stop and synchronous DTEs which, prior to the introduction of the PSPDN, had been used in CSPDNs, PSTNs and on leased circuits.

In the case of direct access of a DTE to a PDN, transmission media utilization economy can be achieved by using statistical time division multiplex systems. They provide the possibility of allocating access circuits to data stations only when they are actually needed for data transmission.

The repertoire of data transmission services provided by PDNs (Table 2.1) has been chosen to make the PDN competitive with other telecommunication networks (Telex, PSTN, including leased circuit facilities) in this domain. This applies not only to these networks themselves but also to their transmission facilities, for example, the standard telephone channel and the primary group channel digitized by modems, and digital channels available in voice frequency telegraphy (VFT) and in pulse code modulation (PCM) systems.

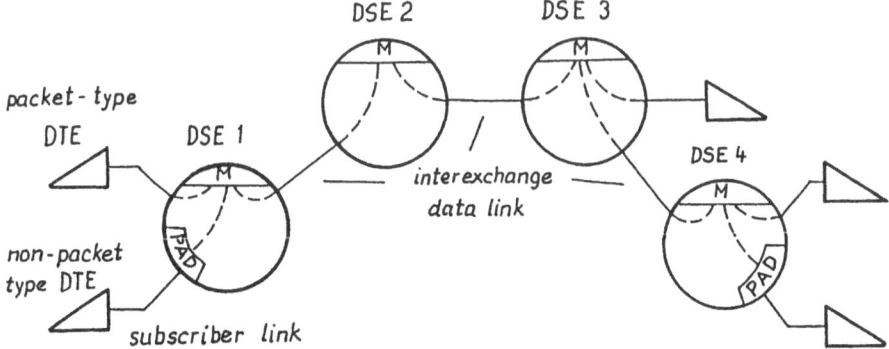

Fig. 2.7. The principle of a packet switched PDN. DSE, date switching exchange; M, memory; PAD, packet assembler/disassembler.

2.3 Data Transmission in the ISDN [3,30,44,71,78]

The ISDN is a telecommunication network that supports a large range of different telecommunication services by providing digital connections or by transmitting digital signals between user-to-network interfaces. This definition shows that the ISDN is capable of providing data transmission services in a way similar to that of PDNs. Hence a suitably equipped ISDN can play the same role as a CSPDN or a PSPDN, or both these networks. This means that in addition to terminal equipment such as telephones, facsimile equipment and teleprinters, the ISDN is capable of supporting DTEs connected to it either individually or in a multiple-terminal installation. Additionally, the ISDN can serve the PDN as an access network for distant DTEs (as can the PSTN).

These ISDN capabilities are illustrated in Fig. 2.8. Any piece of terminal equipment (TE) is connected to the ISDN via a network termination (NT) unit. Point S is situated on the subscriber side of NT.

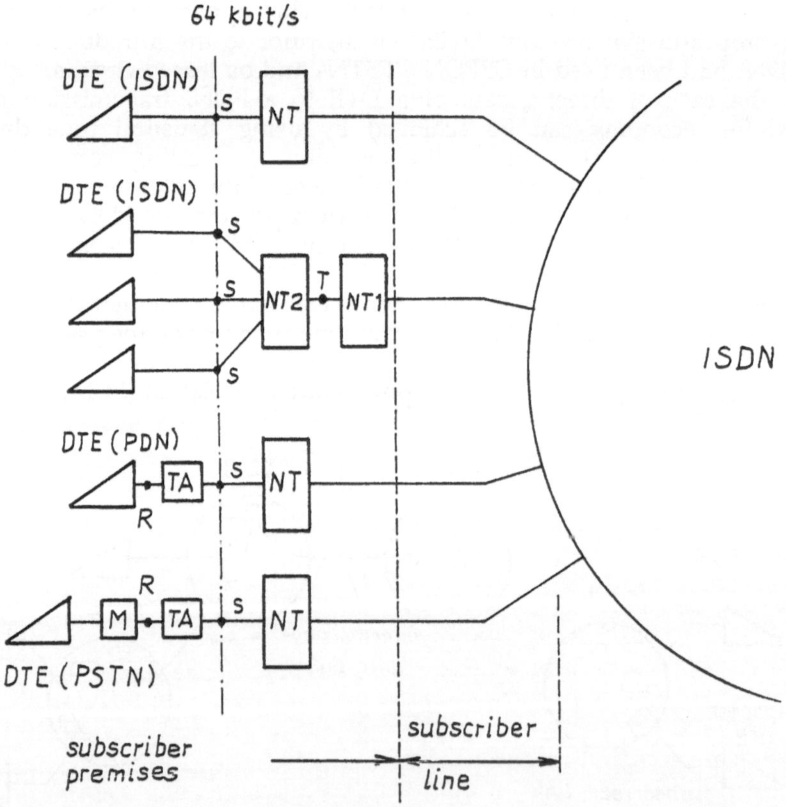

Fig. 2.8. The principle of data communication over the ISDN. M, modem; NT, network terminating unit; TA, terminal adaptor; R,S,T, reference points.

In an installation with more than one TE the NT is split into two parts: NT1 which terminates the subscriber line and NT2, to which ISDN-compatible TEs (TE2s) are connected. Point T is located between NT1 and NT2. The standard rate for an ISDN-compatible TE (TE1) in a narrow-band ISDN is 64 kbit/s (see Sections 3.2 and 4.5.2). If TE1 is a DTE it corresponds to user class 30 in Table 2.1.

DTEs originally designated for use in non-ISDN networks (TE2s) have to be connected to their NTs via terminal adaptors (TAs). In the case of PDN-oriented DTEs they correspond to all the other user classes in Table 2.1. Point R is located between a PDN-oriented DTE and a TA.

One version of TA supports the connection of DTEs primarily intended for connection to the PSTN. In this case the DTE is connected to the TA with interface V.24.

The very principle of the ISDN makes it possible to establish complete switched data circuits between asynchronous or synchronous DTEs, because use can be made of the basic access circuit of the type 2 B_{64} + D_{16} or primary access circuit 23 B_{64} + D_{64} or 30 B_{64} + D_{64} in either direction where B is the basic channel, D the auxiliary channel and the subscripts represent the bit rates in kbit/s. By time division the B-channel or the D-channel can yield subchannels for synchronous DTEs of user classes listed in Table 2.1. A special start–stop-to-synchronous conversion method which can be characterized as "carrying samples resulting from start–stop reception by a synchronous bit stream" enables the implementation of start–stop user classes. Thus the ISDN can fully replace the CSPDN. As such, however, it allows communication only between DTEs of the same user class.

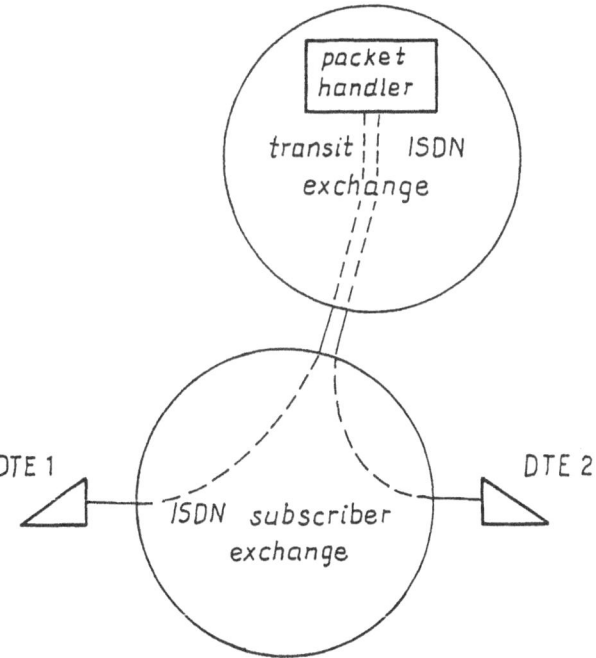

Fig. 2.9. Access of packet oriented DTEs to a packet handler in the ISDN.

With packet switching user classes, the ISDN can either serve as an access network to the PSPDN or perform packet switching by itself. In the latter case an ISDN exchange must be equipped with a packet handler (PH) (Fig. 2.9). A packet-type DTE communicates with the PH to achieve data transmission in the packet mode. An ISDN with exchanges adapted for packet switching can fully replace a PSPDN.

Another possibility of packet switched data transmission over the ISDN is the utilization of the D-channel which, together with signalling system No.7 between ISDN exchanges, supports packet mode transmission (see Section 4.7.3).

3 ■ Switching in Networks

3.1 Network Capacity Sharing [8,14,45,58,70]

Transmission means will always constitute the most expensive part of telecommunication systems and networks and must be handled with the highest possible economy. The problem of how to divide the aggregate transmission capacity and assign its portions to users has existed for a very long time and will always remain, despite many sophisticated solutions supported by advanced technology.

The transmission capacity of a network depends upon the number of circuits within this network, their frequency band, and the time reserved for transmission. It may be represented as a prism or cube in three-dimensional space (Fig. 3.1). Network capacity may be increased by additional circuits, by enlarging bands, and by prolonging time. As the three parameters are

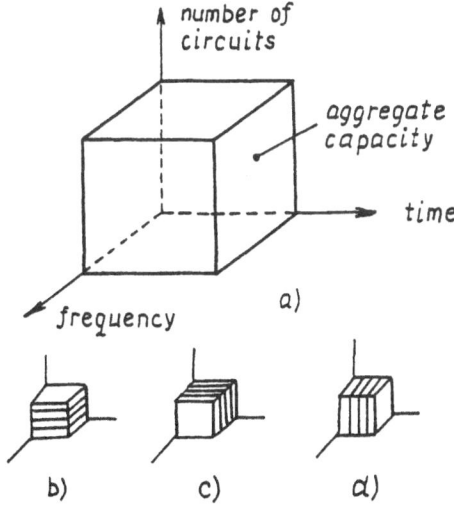

Fig. 3.1. The transmission capacity of a network. **a** Aggregate capacity of a network; **b** space division; **c** frequency division; **d** time division.

mutually independent, three basic methods of capacity division among communicating users are available:

- Space division – a separate circuit is assigned to each communicating pair
- Frequency division – a portion of the circuit frequency band is assigned to a communicating pair for the whole time
- Time division – portions of time are consecutively assigned to communicating pairs employing the overall frequency band capacity

Space division is one of the oldest methods and gave rise to switched networks based upon *circuit* or *line switching*. When two users (network subscribers) want to communicate with each other, the network assigns them a connection composed of circuits (subscriber circuits, trunk circuits) for the time needed for the communication. Such a connection is called a *switched connection* or *switched circuit*, in contrast to the *leased connection* or *leased circuit*.

Circuit switching has progressed through using switching centres (exchanges) of different development generations. The first three generations employed electromechanical and electronic switching fields (hence space switching exchanges); the newest generation, however, uses the time division technique.

Frequency division, also called *frequency division multiplexing* (FDM) is an old, but not obsolete, technique. It has been used in voice frequency telegraph systems. FDM makes it possible to share a common line or channel among several independent signal flows. These flows are separated by appropriate filters on the sending and receiving sides. Each sub-band so constructed is permanently assigned to a signal transmitter–signal receiver pair (see Section 2.1), thus representing a subchannel. The assignment is, as a rule, permanent because it is difficult to apply a mechanism to vary an assignment. FDM thus replaces a bundle of physical circuits wired permanently to assigned pairs of signal transmitters and receivers representing channel inputs and outputs.

Most applications of FDM have been in the area of carrier telephony. Analogue telephone signals modulated with 4 kHz spacing and the primary group of 12 speech channels are fitted into a 48 kHz band positioned between 60 kHz and 108 kHz. This technique is further extended to supergroups of 60 channels, to master groups of 300 channels, and so on, needing, of course, larger and larger frequency bands (10 800 channels require 55.352 MHz band) and special transmission media, such as coaxial cables or microwave radio. FDM is now, however, of less relevance because of the digitization of analogue signals for which *time division multiplexing* (TDM) has become important.

TDM is based upon the fact that there is direct dependence between time and frequency: either a signal changes quickly and requires a broader frequency band, or slowly and requires a narrower frequency band. If the frequency band is divided into narrower sub-bands, the cyclically-repeated time interval frame (do not confuse with the term "frame" in Chapter 4) is also divisible into shorter time slots assigned to independent signals (Fig. 3.2). Similarly, as portions of the frequency band are called frequency division channels, the sequence of consecutive time slots bearing the same serial number within the time frame constitute time division or *logical channels* (the time frame in Fig. 3.2 gives rise to logical channels 1 to 5). Division of time into frames is called *slotted division*.

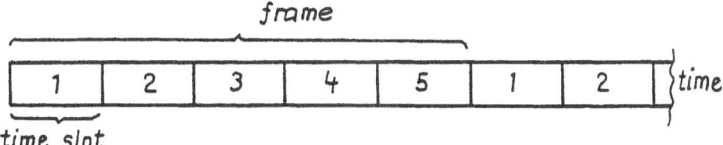

Fig. 3.2. Division of time in frames and slots.

The division (frequency division as well as time division) is the first step in utilizing the capacity of a transmission procedure. The second step consists of the assignment of channels among pairs of communicating stations. As mentioned above, FDM is able to assign frequency channels only statically, that is, permanently to pairs of stations regardless of their status, requirements and traffic intensity. TDM is much more flexible. It allows not only static assignment, as in (static) time division multiplexers but also dynamic assignment.

Fig. 3.2 is in fact an example of static assignment or synchronous time division multiplexing (STDM) which, as with FDM, does not distinguish between busy channels occupied by communicating stations and idle channels allocated to stations temporarily silent. In order to increase capacity employment, the idle channels are occupied by another pair of stations which wants to communicate at that time. Fig. 3.3 shows an example of five logical channels allocated among eight pairs of stations, a to h, five of which require transmission capacity (in the first frame pairs a, c, e, g and h communicate, in the second frame pairs e and g become silent and are replaced by pairs d and f). Such an assignment, known as statistical, demand, or asynchronous time division multiplexing (ATDM), was proposed for data communication in 1969 although it was in operation ten years earlier with the aim of increasing the capacity of the Atlantic Ocean cable for analogue speech transmission (time assignment speech interpolation, TASI).

Though dynamic assignment allows an increase in the number of communicating stations without extending the time frame, the penalty for this is an increase in waiting times and hence the overall response time. In addition, each logical channel, or even each time slot, needs to be designated with addresses of stations, which occupies slot space assigned originally for user information.

Nevertheless, dynamic assignment is preferred to static assignment, particularly in networks with conversational terminals whose activity does not exceed a small percentage of the time. It is not surprising that this technique has also been applied with certain modifications in radio and satellite terminal networks (time division multiple access, TDMA; slotted or S-ALOHA; carrier-sense multiple access, CSMA) and, later on, in LANs (carrier-sense multiple access with collision detection, CSMA/CD, token passing).

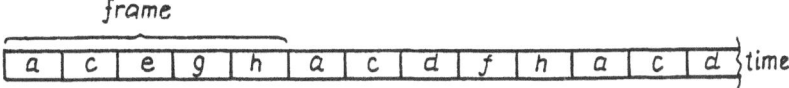

Fig. 3.3. An example of dynamic assignment.

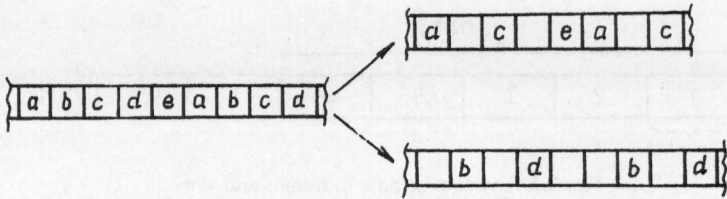

Fig. 3.4. Circuit switching by means of time division multiplexing.

TDM is applicable to digital signals only. Digitization of analogue signals has increased its value substantially. Voice signals are carried digitally by pulse code modulation (PCM) employing 64 kbit/s for each voice channel. The signal is first sampled 8000 times per second and each analogue sample is then coded into 8 bits, resulting in a stream of 64 kbit/s. Multiplexing onto a sufficiently high signalling rate allows the interleaving of several independent coded voice channels in one sampling period of 125 µs. For example, 32 channels require 2.048 Mbit/s, 128 channels require 8.448 Mbit/s, and so on. This creates the hierarchy of PCM transmission multiplexing schemes up to, at present, 564.992 Mbit/s or 8192 speech channels, which is more efficient than conventional carrier telephony transmission systems.

Time division, as well as frequency division, was first applied to concentrating the information flows from several stations onto one transmission medium. This method, however, also allows the separation of individual channels from the common signal stream along the transmission path, and their independent routing. Fig. 3.4 illustrates an example of separating five channels from the common time-multiplexed stream and their routing in two different directions. If the separating point is placed in a network, it is in fact a switching node and the demultiplexing plays the role of switching. This forms the basis for time division exchanges or PCM exchanges (fourth generation exchanges) and for integrated digital networks (IDNs), as well as for integrated services digital networks (ISDNs).

Up to now we have dealt with slotted multiplexing techniques but the stream of time slots need not be driven by a fixed clock. Fig. 3.5 shows the unslotted concentration from three independent sources hosted in stations 1, 2 and 3. No frames are created on the common path even though three logical channels are distinguished by their first numbering digit. The function of division and assignment is called *concentration* and is performed by *concentrators*.

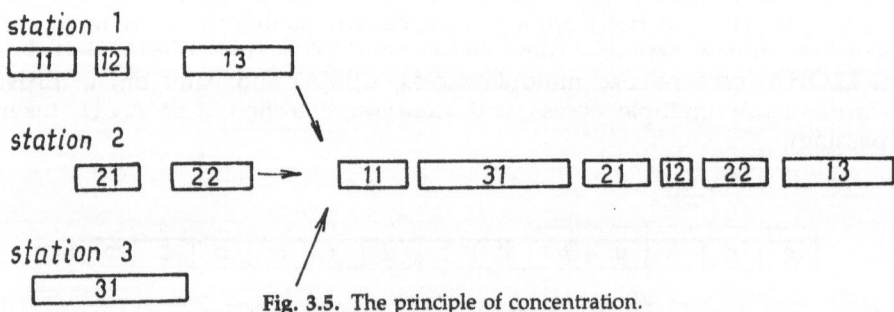

Fig. 3.5. The principle of concentration.

Concentration is very similar to ATDM. However, it gives longer time delays if messages are long, and no priority or preference scheme is applied. This may be seen in the case of station 2 in Fig. 3.5, which has to wait even with a very short message until station 3 completes the sending of its first message.

To relieve stations from controlling the status of a common path and overall traffic, the concentrator is equipped with buffers for individual messages as well as with longer-term storage. The buffers enable the creation of individual queues at the low-speed lines to and from stations; the long-term storage forms a queue at the higher-speed line. The whole control of concentration is performed within the concentrator proper and stations may only demand preferences and other facilities if they are provided by the concentrator. Unslotted time division and assignment need not always be controlled. For example, the unslotted multiple access "pure ALOHA" (P-ALOHA) is a random, uncontrolled method. This system was designed and tested in the Hawaiian network of mobile stations in 1970 and has been successfully applied in contemporary very small aperture terminals (VSATs).

Concentration is a store-and-forward technique. Messages taken from stations are temporarily stored and later forwarded to one outgoing route. Conversely, incoming messages are stored and then routed to stations according to addresses located in message headers. If no difference between outgoing and incoming messages is made, a concentrator becomes a message switching node. Message switching could not develop within the PSTN since telephone conversations require real-time interaction. Within telegraph networks, however, messages should be delivered accurately, and transmission delay is not of vital importance. Message switching has been used in telegraph networks only since 1940 (for example, the AT&T torn-tape exchange 81-A-1).

Message switching has many advantages over circuit switching. Messages are stored for possible later search (in case of their loss), and they are accepted by the network regardless of the state of the addressee. Error control may be performed within the network and the speed (data signalling rate), alphabets and codes can be changed. This allows completely different DTEs to communicate with each other. Messages may be held within the network and made available to the addressee when explicitly released by the sender (quarantine service).

On the other hand, as well as the exclusion of real-time interaction, message switching requires limited message lengths, and possible network congestion further increases the time of message delivery. Circuit switching provides transparent physical connections with negligible transmission delays (the substantial delay is caused during the connection establishment phase, which may also be increased as a result of network congestion).

Figs 3.6a and b compare the two methods of switching. Examples are shown on a very simple network consisting of one node, over which two stations (DTEs) communicate. Note that circuit switching requires addressing only for connection establishment, while in message switching each message has to be addressed since the transmission path is determined by the address.

Although the store-and-forward methods have an irreplaceable role in data communication, a way of reducing delays needed to be found. Fig. 3.5 shows that considerable delays are caused by long messages and it therefore became

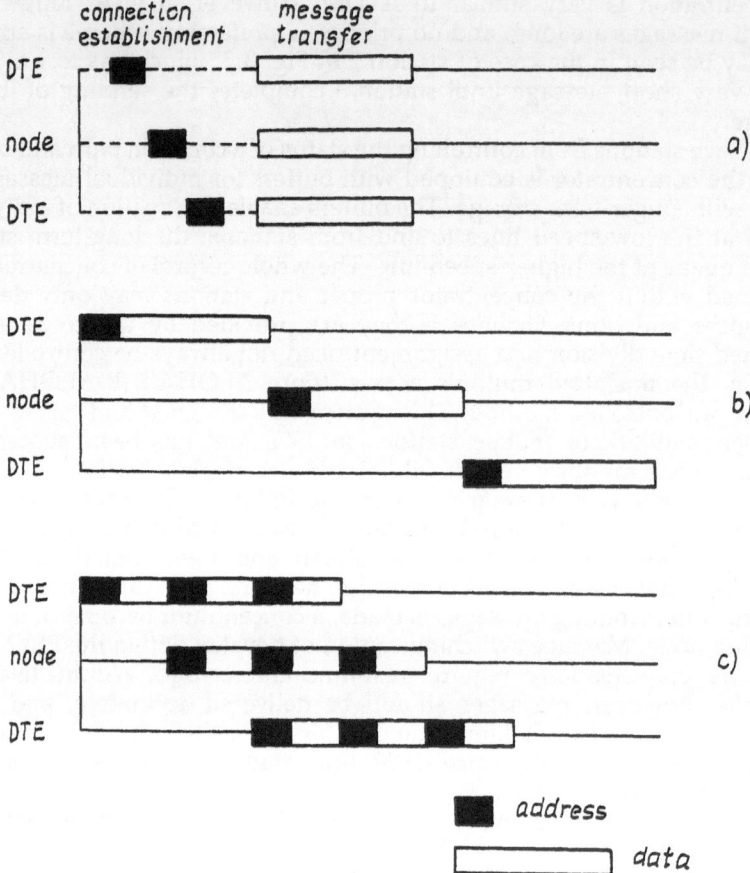

Fig. 3.6. Methods of switching. **a** Circuit switching; **b** message switching; **c** packet switching. Note, the transmission and storing delays are neglected in the three examples.

evident that a strictly limited size had to be imposed on messages. Shortening of messages, however, cannot constrain users from their applications and thus the breaking of messages into fragments has to be admitted without padding out shorter fragments to the maximum size. Fragments with addresses and other necessary control and identification information form packets and hence the need for *packet switching* became evident.

Fig. 3.6c illustrates an example of delivery of a message broken into three fragments over a packet switching node. For the purpose of comparing the three methods of switching the same size of message is illustrated. The figure proves the advantages of packet switching over circuit switching and message switching in the case of short messages.

Packet switching, in the same way as message switching and concentration, implies the interleaving of packets on lines. Packets with the same number, which refer to the same message, constitute, as in other time division techniques, logical channels. Fig. 3.7 shows an example of three

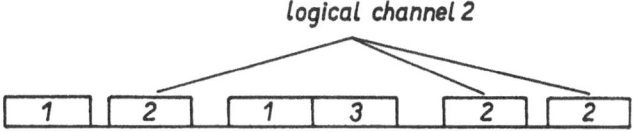

Fig. 3.7. An example of logical channels.

logical channels numbered 1, 2 and 3: logical channel 1 is made up of two packets, logical channel 2 of three packets, and logical channel 3 of only one packet.

Table 3.1 compares the features of the three switching methods. Note that all of the features of message switching are retained in packet switching, and the delays are reduced to less than 1 s, allowing a sufficiently rapid exchange of information resulting in quasi-conversation. The areas of application appropriate to each switching method are described in the last row of the table. From this it follows that no switching method is preferable to any other.

Table 3.1. Comparison of switching methods

Circuit switching	Message switching	Packet switching
Physical connections	Virtual connections	Virtual connections
Real-time interaction (conversation) possible	Real-time interaction impossible	Quasi-conversation possible
No storing	Messages are stored for possible later searching	Packets are temporarily stored during transfer
Connection established for the whole communication	Transmission path is determined specially for each message	Transmission path may be determined for each packet
No connection is established if the addressee is busy	Message is accepted regardless of the state of the addressee	Packet is detained or returned to the sender if the addressee is busy
Delay caused by the connection establishment	Substantial delay caused by storing messages	Delay less than 1 s for one packet transfer
Error control is a matter of user decision	Error control partially performed by network	Error control completely performed by network
Network congestion prevents connection establishment, no influence during data exchange	Network congestion increases the time of message delivery	Network congestion stops packets penetrating into a network
No restrictions upon length of data message	Message lengths are limited	Data messages must be cut into length-limited fragments either by user or by network
Connection is speed- and code-transparent	Speed and code conversion possible	Speed conversion usual, code conversion possible
Quarantine service impossible	Quarantine service possible	Quarantine service and redirection of calls possible
Efficient for small traffic intensity and long data messages	Efficient for medium traffic intensity and data messages of limited lengths	Efficient for large traffic intensity and short data messages

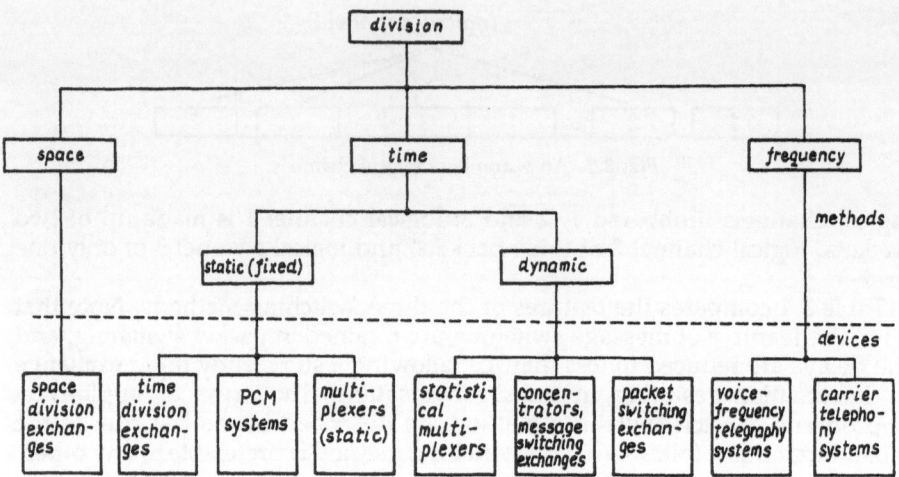

Fig. 3.8. Methods and devices of transmission capacity sharing.

However, circuit switching and packet switching became the basis for public switched networks, and both types of network often operate side by side, giving the users the option of choosing the method best suited to their needs.

A summary of methods of transmission division and types of devices employing them is illustrated in Fig. 3.8. In Sections 3.2 and 3.3 we shall examine circuit switching and packet switching in more detail, because of their importance for PDNs as well as for ISDNs.

3.2 Circuit Switching [1,6,8,22,42,69]

Circuit switching in PDNs is based upon the rich experience gained from the operation of automatic switching systems in the telephone network since the beginning of the twentieth century and in the teleprinter exchange (Telex) network since the 1930s. By definition, a circuit switched network is a telecommunication network providing telecommunication services based upon the use of temporarily-established circuits for the transmission of signals between telecommunication terminal equipment: telephones, tele-printers, and – in the case of CSPDNs – DTEs.

In contrast to conventional (interpersonal) circuit switched telecommunication networks the term DTE or user is often used instead of "subscriber" for a communication participant, because in many cases human beings play no part in the communication; for example, when the call is set up by the computer or when the terminal sends a data message at a predetermined time, in the absence of the operator.

The germ of the CSPDN can be traced to the endeavours to use the Telex network for data transmission (CCITT recommendations of the S series), to connect the DTE to the Telex network and to use the telephone network for data transmission (CCITT recommendations of the V series, the first of which

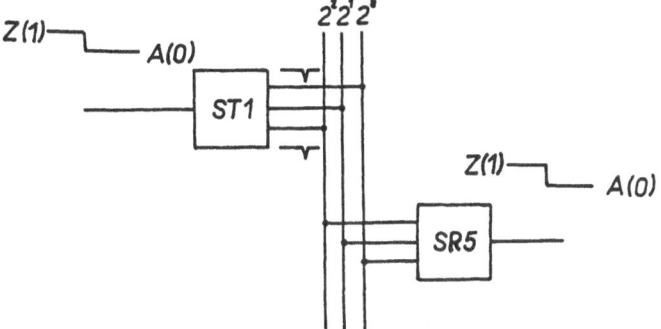

Fig. 3.9. The principle of the transmission of significant instants through a CSPDN.

appeared as early as 1960). The opportunity to establish networks specializing in data transmission was given to telecommunication administrations at a time when the market brought digital switching systems capable of integrating the services of Telex and data transmission. Beside this integration, the integration of switching and transmission techniques is another essential feature enhancing the cost-effectiveness and reliability of modern exchanges. The introduction of multichannel transmission systems on the basis of TDM paved the way for this integration. The development of switching systems for CSPDNs was facilitated by the application of experience gained with the design of PCM-based digital telephone switches.

The basic function of an exchange in a CSPDN is the same as that of an exchange in the telephone or teleprinter network – to establish a switched circuit for communication between a calling and a called subscriber. One possibility is the address switching principle used in early data circuit switching for the transfer of significant instants: instants of change from condition A(0) to condition Z(1) or vice versa of the binary signal through the switching field realized by a common bus of parallel wires, as shown in Fig. 3.9. During call set-up the input signal transmitter ST1 registers 5, the number of the output signal receiver (for SR5). Each time a change occurs at the input of ST1, this transmitter sends pulses representing the number 5 in binary notation (101) to the parallel bus. Only receiver SR5 evaluates this combination by changing the binary condition at its output. The system has to be protected against collisions by preventing simultaneous passages through the bus. This example, together with Fig. 2.1, demonstrates that a temporary channel terminated by ST1 and SR5 has been established within the exchange during call set-up (and will cease to exist after the call has been cleared by removing 101 from ST1).

The use of a common memory for the same purpose is illustrated in Fig. 3.10. During call set-up the exchange connects circuits 7 and 19 (input 7 with output 19 and input 19 with output 7) by storing 19 in cell 7 and 7 in cell 19. A change arriving at input 7 addresses (via the input code converter in binary) cell 7 where 19 has been stored since call set-up. Cell 7 sends the stored address in binary form to the output code converter. This converter addresses the output corresponding to the stored number (output 19). The same method can be applied to signal samples or whole characters.

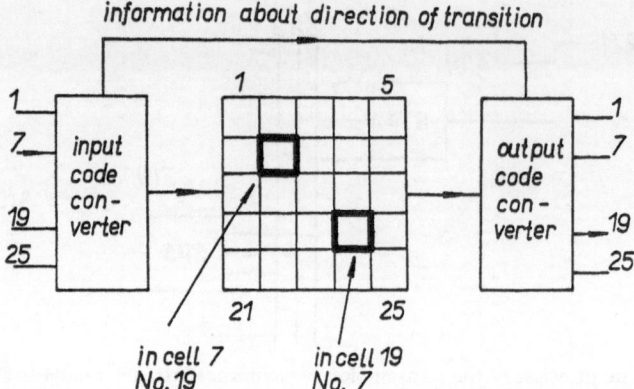

Fig. 3.10. The use of a common memory for allocation of outputs to inputs in a circuit switching digital exchange.

The development of microelectronics and computer techniques created favourable conditions for harnessing the time division principle (see Sections 3.1 and 4.5) for transmission as well as for switching. In fact, a time switch can be regarded as a TDM system in which the distance between multiplexer and demultiplexer is negligibly short and output time-slot assignments are subject to the required connection of incoming and outgoing circuits in the chains between communicating DTEs.

The transmitted bit streams can be either those generated in the DTE or samples of the signal to be forwarded through the exchange. A signal presented for switching is usually a binary signal. If the sampling frequency is higher than the bit rate of the incoming data signal, the channel through the exchange is code- and speed-transparent; that is, it can carry any binary signal not exceeding a given limit of modulation rate or – in other words – having a certain minimum unit interval.

The storage of a sample in memory and its transmission at a later point in time causes an undesirable transmission delay but is necessary to strip the signal of telegraph distortion; that is, to regenerate it. This applies particularly to the user classes of start–stop transmission. Table 3.2 summarizes the switching principles commonly used in CSPDNs.

Table 3.2.   Digital circuit switching methods

Principal classification	Principle	Transparence	Regeneration
Space switching	Contact switching field	Yes	No
	Semiconductor matrix	Yes	No
Time switching	Multiple sampling	Yes	No
	Transfer of data bits	No	Yes
	Transfer of characters	No	Yes
Address switching	Transfer of significant instant	Yes	No
	Transfer of data bits	No	Yes
	Transfer of characters	No	Yes

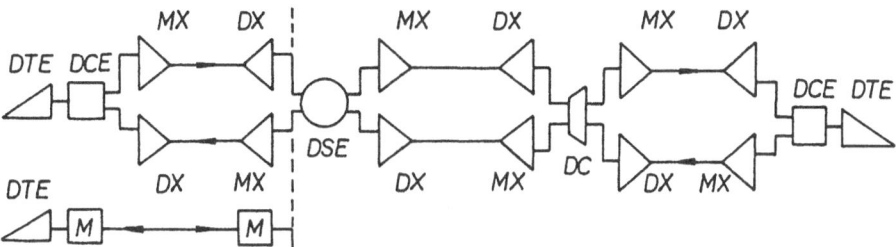

Fig. 3.11. The means of transmission involved in DTE-to-DTE transmission in a CSPDN. DC, data concentrator; DSE, data switching exchange; DX, demultiplexer; M, modem; MX, multiplexer.

CSPDN exchanges work with stored program control. During call set-up the DTE communicates with its parent exchange or concentrator (which plays the role of an exchange though with limited functions, the full assortment of these being performed by the above, fully equipped exchange), utilizing specified signals and procedures within the subscriber signalling system defined by CCITT recommendations (see Section 4.7.2). To extend the call to the called subscriber a similar communication must take place between the exchange and another exchange or the called subscriber, depending on which exchange in the network this subscriber belongs to. These individual communications result in the establishment of an end-to-end connection between the calling and called DTEs over a complete data circuit, enabling the exchange of data.

In the call set-up phase the DTE communicates with the exchange in the nominal bit rate of the corresponding user class of service and in the IA5 code. In addition, start–stop classes have a fixed format for the start–stop character.

It is characteristic of the circuit switching system that every link (elementary circuit) of the chain constitutes a complete end-to-end circuit and preserves its data signalling rate. It consists of two channels of opposite directions implemented in multichannel systems based on the principle of time division (TDM). Subscriber–exchange DTEs are as a rule connected directly (individually) by means of modems or baseband signal converters, or, in the case of subscriber concentration around a certain point or along a certain line, by time division or even frequency division multiplex systems. These multichannel systems are applied whenever a concentrator is not an economic solution.

The structure of a switched end-to-end circuit in a CSPDN is shown in Fig. 3.11. The subscriber circuit can be individual or part of a multiplex system with time division or, rarely, frequency division. The interexchange circuits are realized solely in TDM systems. The role of multiplex systems in a CSPDN is evident from Fig. 3.12.

Circuit switching for circuit switched data services in the ISDN is done on the basis of the 64 kbit/s bit stream transmission for user class 30 according to CCITT recommendation X.1, which is equivalent to the 64 kbit/s class 19 provided by the CSPDN. The provision of all other CSPDN user classes, that is, those of lower bit rates than 64 kbit/s, is done via terminal adaptors (TAs), whose main function is to insert the lower data rate data transmission bit stream at reference point R into the 64 kbit/s bit stream at reference point S/T. The principle of this insertion is a matter of multiplexing and will be explained in Section 4.5.3.

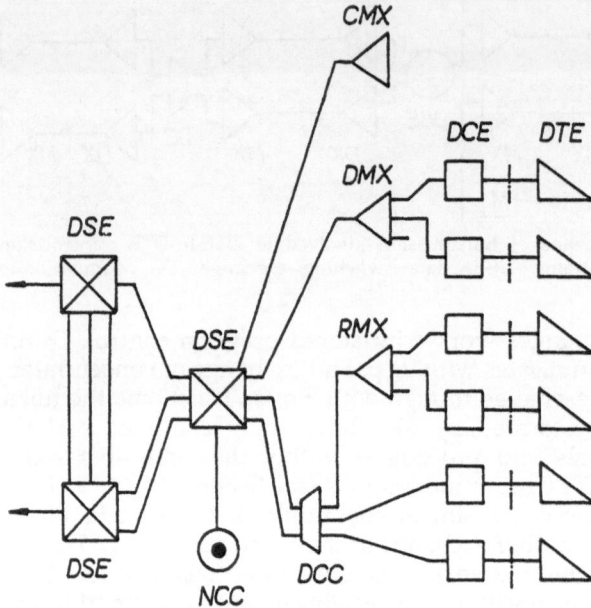

Fig. 3.12. The role of multiplex systems in a CSPDN. CMX, private (company-owned) muldex; DCC, data concentrator; DSE, date switching exchange; DMX, data muldex; NCC, network control centre (see Section 3.4); RMX, remote data muldex.

3.3 Packet Switching [5,13,14,22,42,47,51,53,72]

As we have seen in Section 3.1, packet switching is in fact not a new invention but a reapplication of earlier TDM and dynamic assignment. The first published description of what is now called packet switching was in a large eleven-volume study "On distributed communication networks" prepared by P. Baran of the Rand Corporation in 1964 (a survey has been published in [2]). This study proposed a fully-distributed packet switching system (not calling it so) for voice and data communication including a digital microwave transmission and security capability. The United States Air Force as a customer, however, did not continue this project and the results had to wait until packet switching was rediscovered and applied by others.

Further development of packet switching was carried out by two pioneers: D. Davies of the National Physical Laboratory in the UK and L. Roberts, at that time at the Massachussetts Institute of Technology in the US. The former proposed a public digital communication network based upon a 1.5 Mbit/s PCM transmission system equipped with assemblers/distributors, the predecessors of the packet assembly/disassembly facilities. The latter participated in the design of the Advanced Research Project Agency – the ARPA network which was planned to link host computers at leading universities and research laboratories by means of packet switches and

50 kbit/s data circuits. However, the term "packet" was coined by Davies to name a block of 128 octets being moved about and treated as a whole. The first published documents on these projects were coincidentally presented in October 1967 at the ACM Symposium at Gatlinburg (USA) [15,61].

Nevertheless, the first public demonstration of packet switching took place five years later during the first International Conference on Computer Communication (ICCC) in Washington. A complete ARPA packet switching node was installed at the conference hotel, with about forty terminals permitting access to many remote host computers. This event convinced the majority of conference attenders not only of the vitality of this technique but also of the fact that a large computer network composed of over one hundred elements (host computers, node minicomputers, wideband circuits) is able to function extremely reliably. The development of packet switching, both for the user and public data networking, has since accelerated.

Section 3.1 briefly described the principle of packet switching and compared it with the other types, circuit switching and message switching. In this section packet switching and corresponding services are dealt with in more detail.

While in circuit switching data messages are transmitted from a source to a destination irrespective of their lengths, packet switching either limits message lengths or, if the length exceeds the permitted size, requires fragmentation. Examples of the former case include interactive communication, database queries and electronic funds transfer. Short data messages are enveloped into independent packets bearing the addresses of their destination and origination and put into a packet switched network. Packet switching network nodes route packets to demanded directions in order to reach the right destination as quickly as possible, but reliably. The demand for speed implies that the shortest paths will be chosen, whereas the reliability aspect requires failed circuits to be bypassed at the cost of path elongation. It is of no importance whether the packets reach their destination in an order different from that in which they were launched.

Fig. 3.13 shows part of a packet switched network with two switching nodes and four users (one host and three terminals labelled a, b and c). Short data messages conveyed as a whole in individual packets are exchanged between the host and the three terminals in both directions. Consecutive

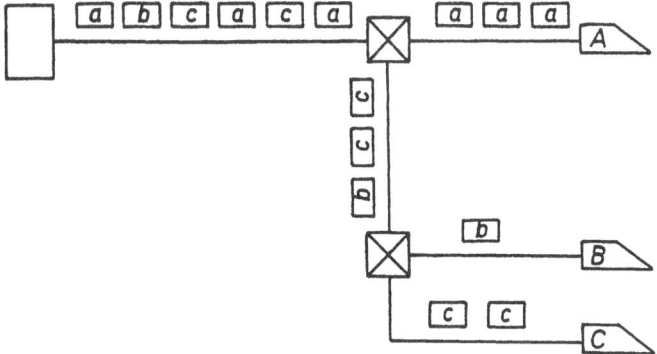

Fig. 3.13. Part of a real packet switched network.

packets carrying the same labels form logical channels between user and switching centre and between two centres. When a packet enters a switching centre, it is stored, the destination name (address) is read and the packet is routed to the queuing buffer in the determined direction, where it awaits release. Packets of the same logical channels may be routed in different directions according to the load in outputs.

A problem arises if data messages are too long to fit into the packet data field. Then the message has to be split into fragments and each fragment has to be enveloped into a separate packet. However, even though packets are launched in the right order following the fragmentation of the data message, the packet switched network cannot always guarantee the same order of packet arrival, because packets are handled independently. Some packets belonging to the same message may outrun others, some are delayed due, for example, to retransmissions. It is up to the addressees to reassemble received packets according to the original data message. Fig. 3.14 shows the necessary actions to be taken at the sending and receiving sides. Each fragment has a header containing addresses, control information (data), identifiers, parameter values, etc. (see Section 4.7.3), which facilitates packet handling within the network and is not removed until the packet reaches its destination.

Most private packet switched networks operate on the principle that the reassembly of the fragments is the responsibility of the user. However, the development of PDNs forced their designers into attempting to overcome this obstacle, because preservation of the order of packets is a feature of data transmission over leased or switched circuits and is more user friendly. These attempts eventually resulted in the *virtual circuit*, a concept already outlined in Rand Corporation studies.

The term "virtual" was taken from the Latin word "virtus"(ability, capacity, power, virtue) and is used to describe an imaginary object possessing the same properties as a real one. In circuit switched networks the transmission path is established by the interconnection of different circuits (subscriber, junction, trunk) done by exchanges. During the call establishment phase both stations and the network can agree types of services and their quality (performance), including costs and user facilities. Of course, both switched and leased circuits preserve the correct sequence of information units as launched by the sending station.

Let us return to packet switching and try to imitate the circuit switched or leased circuit connection. The first packet opening the communication is

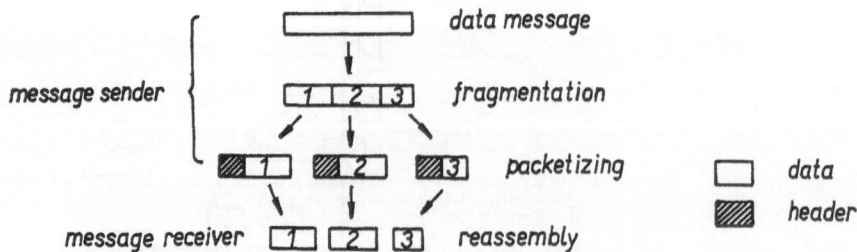

Fig. 3.14. Data message fragmentation, packetizing and reassembly.

launched into the network. The switching node finds a storage location for it and determines the direction of its next journey. The storage location need not be released to serve arbitrary packets but can remain pre-allocated to the same logical channel (that is, for packets of the same correspondence). After the first packet reaches the destination (addressee), a tandem of logical channels joined by fixed storage locations ensues between sender and receiver. Fig. 3.15, which corresponds to the layout in Fig. 3.13, shows three virtual circuits composed of logical channels between the host and the terminals. Virtual circuits are labelled by lower case letters referring to DTEs, which are designated by upper case letters, and the interconnections of logical channels are represented in switching nodes by dotted lines in order to distinguish them from physical interconnections. Virtual circuits permit two-way simultaneous communication. They preserve the correct sequence of all packets in the process of communication, including packets where long data messages are broken up into fragments. However, they retain all the characteristics of packet switching (accommodation to data signalling rates of different DTEs). In addition, they increase performance, particularly regarding residual error rate and throughput. Each logical channel operates individually; that is, error control is performed within shorter sections (this feature is inherent to packet switching). Subsequent packets need not include the address because their path has already been traced out by the first packet. This is important because if this were not the case the packet overheads would reduce efficiency and, as a result, increase costs. Reliability seems to have decreased because of loss of routing flexibility but the network is capable of reconnecting logical channels when a line failure occurs. If packets are sequentially numbered, the reconnection can be accomplished without loss of packets as quickly as packet rerouting in the case when no virtual circuits could be established. Virtual circuits have well-defined properties and the quality can be agreed before their use as in circuit switching. Last but not least, addressees are relieved of reassembling the fragments.

Therefore two basic services are provided by the packet switched network:

- Virtual circuit – VC (or virtual call or connection-mode) service
- Datagram – DG (or connectionless (CNL) mode) service

VC was described above. It requires the establishment (or virtual call set-up) of the exchange of ancillary packets containing no user data. An example of virtual call set-up is shown in Fig. 3.16. The initiating DCE sends the packet

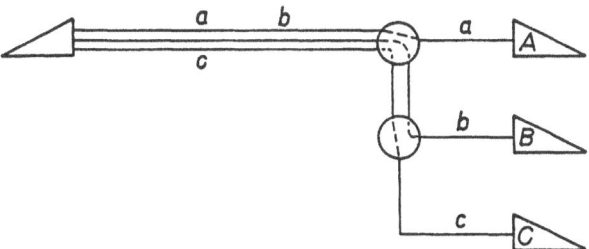

Fig. 3.15. The concept of virtual connections.

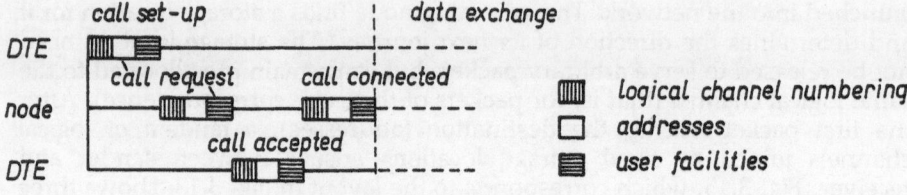

Fig. 3.16. Virtual call set-up.

"call request" (the names of packets are taken from PSPDNs), which appears at the called DTE as the "incoming call" packet (not shown in the figure). Tracing packets are usually shorter than data packets. They contain the designation of logical channels used, the addresses of the two stations (calling as well as called) and demands/confirmations of user facilities and quality values. The called DTE reacts by sending the packet "call accepted" if it wants, or is able, to be connected and the network informs the calling DTE by the packet "call connected". From this instant the virtual circuit is established and ready to serve the exchange of packets carrying only user data fragments (without addresses, commands and identifiers).

Note the duration of the call set-up phase and compare it with Fig. 3.6. It is clear that no benefit is gained by virtual circuits if data messages are short and contained in single packets. On the other hand, the transfer of long messages profits by the use of virtual circuits because the call set-up phase is negligible compared with the duration of the data transfer phase.

The call set-up phase can be eliminated in the case of leased lines provided by the network (note that the same applies to circuit switched networks). If agreements between the two stations and with the network about user facilities and performance values have been made in advance, a virtual circuit may be established for a contractual period of time and hence plays the role of a leased physical line. This results in the so-called *permanent virtual circuit* (PVC) service, to be distinguished from the switched virtual circuit (SVC) service or virtual circuit (VC) service. Except for virtual call set-up and release phases, there is no difference between the two services (for comparison see Table 3.3).

Network designers and users have not forgotten the original packet switching principle of individual transport or self-contained packets. Such a service differs substantially from the VC service from both the network provider's and network user's points of view: (i) no virtual connection is established, which implies no possibility of making an agreement about the necessary parameters; (ii) storage allocation and routing are accomplished, as required, for each packet individually, meaning that each packet needs to have an address; (iii) sequencing of the received fragments is the user's responsibility; and (iv) no information about complete data message delivery is returned to the message sender (the VC service usually provides such an acknowledgement). For details see Table 3.3. The service providing for the transfer of self-contained packets is called the *datagram* service (its etymology being analogous to that of the term "telegram") or *connectionless-mode* service (since no connection is established). It is efficient for short data messages in interactive applications. An example of the datagram service in a packet switched network with three switching nodes is shown in Fig. 3.17. Suppose

Table 3.3. Comparison of packet switched services

Permanent virtual circuits (PVC)	Switched virtual circuits (SVC)	Connectionless datagram (DG)
Virtual connection is established in advance	Virtual connection must be established before each communication	No virtual connection is established
No packet need be addressed	Addressing is needed only during connection establishment	Each packet must be addressed
Routing is set up in advance	Routing is set up during connection establishment	Each packet is routed independently
Node-to-node as well as end-to-end flow control is performed	Node-to-node as well as end-to-end flow control is performed	Only node-to-node flow control is performed
The order of packets is preserved (sequencing)	The order of packets is preserved (sequencing)	Packets are delivered regardless of the order of sending (no sequencing)
Performance values and user facilities are agreed in advance	DTEs may agree the performance values and user facilities during connection establishment	Performance values and user facilities cannot be agreed
Packets and message delivery can be acknowledged by the receiving DTE	Packets and message delivery can be acknowledged by the receiving DTE	Message delivery remains unacknowledged, acknowledgement of separate packets possible

DTE A wants to dispatch a data message which loads into three packets (1, 2 and 3) to DTE B. With the aim of the quickest possible message delivery, the network employs all paths leading to the addressee. Two packets (1 and 3) are routed directly over one data circuit while packet 2 is bypassed over the third node. At the destination node the packets (and hence the fragments) appear in the wrong sequence (1, 3, 2) and DTE B must rearrange them.

The independent routing of each datagram may result in a packet remaining in the network for a long time, thus increasing network congestion. To avoid long packet journeys within the network a lifetime control function was introduced. This may be defined as the maximum

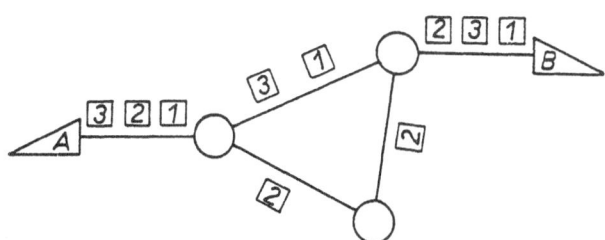

Fig. 3.17. An example of datagram (connectionless-mode) service.

number of hops (nodes permitted to be passed) set by the sender and hidden in the packet header, which is decreased by one at each node. If the lifetime value is exhausted (that is, is equal to zero), the packet is automatically discarded and the sender is informed about the event.

Packet switched networks claim many benefits to users in comparison with circuit switched networks. There is, however, one inconvenience that could discourage users with simple input/output devices from employing the packet switching technique.

It has to be understood that each DTE is equipped with a packet handler which packs packets from bit or character strings and conversely unpacks packets to pick up the original data content. This is easy if DTEs are programmable, such as hosts, intelligent terminals and personal computers. Simple start–stop terminals (teleprinters, visual display units), widely used for conversational systems, seem to be excluded from packet switching benefits. Fortunately, PSPDNs enable access to this type of DTE, through the so-called *packet assembly/disassembly* (PAD) facility, which is simply a packet handler combined, if necessary, with a multiplexer/concentrator to enable a large number of DTEs to make use of it.

The PAD facility collaborates at the DTE side with character-oriented start–stop equipment by exchanging character strings and, at the same time, works with network switching centres by exchanging packets.

3.4 Network Management [4,9,31,43,63,68,77]

Packet switched networks, like circuit switched networks, are extremely complicated and require distributed control. The distribution concerns not only the layout (DTEs, switching nodes, PADs) but also a certain hierarchy of layers within all network elements (see Chapter 4). Although distributed network control is able to cope with almost all situations and events, even in a changing environment, some activities are still missing: overall co-ordination of controlling and controlled elements, monitoring of network state (faults, failures, performance values, traffic), accounting and billing services provided by the network, and, if need be, the protection of user data and user access to network resources. Facilities performing these activities are known as "network management".

The highest level of mechanism for network management is generally in the hands of human operators (hence the term *management*), who employ the management information gathered from network elements through a workstation specializing in the operation, maintenance and administration of the network.

Five categories of management are recognized in networks:

- Fault management – reporting the occurrence and location of faults, and scheduling and reporting diagnostic tests
- Configuration and name management – collecting and disseminating data about the current state of the network and its resources, modifying network attributes and changing network configuration during installation, extension and reduction

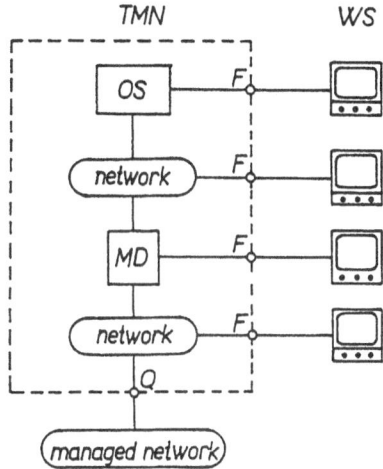

Fig. 3.18. Scheme of telecommunication management network (TMN). OS, operating system; MD, mediation device; WS, workstation; Q,F, interfaces.

- Performance management – collecting and disseminating data about the current network performance level, supervising traffic and changing network configuration to cope with extraordinary traffic conditions
- Accounting management – informing users of costs incurred and collecting billing information
- Security management – providing support for authorization facilities, access control, encryption and authentication

These activities may be accomplished at and co-ordinated from one centre, which is called the network control centre (NCC, in ARPA terminology), network diagnostic centre, network management centre or network administration centre. This relates to medium-scale networks such as national PDNs and private networks. In large-scale networks more than one NCC is preferred, and for still larger networks the telecommunication management network (TMN) has been introduced. This network operates as an overlay network serving for exchange of management information to and from the managed network(s) and managing personal activities at different types of workstations. A telecommunication management network comprises communication networks (for long-distance as well as for short-range transmission), and the necessary operations systems and mediation devices (see Fig. 3.18). Managed networks are connected to the TMN via well-defined interfaces (Q-interfaces in CCITT terminology). Operators' workstations operate through unified user interfaces (F-interfaces).

4 ■ Network Architectures

4.1 Functions and Services [23,52,56,58,79]

A communication process should not be unrestrained, but fully controlled because it serves goal-seeking human or intelligent machine activities. Communication control must cope with many influences, usually random rather than deterministic, and should finally result in a prescribed smoothed behaviour. For example, the impact of noise in data circuits, causing errors, can be stifled by an appropriate error detection and recovery mechanism which, regardless of the stochastic character of influences, will keep the residual error rate within limits prescribed in advance by the users' requirements. As most data communication systems operate in the absence of intelligent human beings, the structure of their control has to be sophisticated. This fact affects the design, implementation and operation of communication systems.

In order to avoid the greatest obstacles a certain decomposition of such a complex system has to take place. This decomposition, however, must follow well-defined rules and conventions which, once set up, will remain permanently in force. There were formerly several ways of decomposing a communication system or its control, but layered decomposition (Section 4.3) is at present supported by many standardization documents and favoured by many microelectronic chip producers and software engineers.

The commonly-used term *architecture* has been chosen to underline the conventional form of organization of communication control. This form is expressed by a structure and rules creating a frame within which the control system is, or should be, built up.

There are several foundation stones of computer/communication systems architecture. However, we begin with two central terms, *functions* and *services*, because they are often confused with their common interpretations, particularly in the context of telecommunication functions and services.

The behaviour of each system (not only the communication control system) is composed of activities performed by system elements or entities. The activity is a response of the system not only to user demands but also to changes of system environment and of the system itself (or part of it). If the user of a communication system fails or refuses to communicate, despite making a request to do so, or if a data circuit or node computer of the PDN degrades its operation, the communication control should start to perform an activity that will lead to another state adequate for the original requirements.

The adequacy is evaluated by a system user who knows, or at least should know, the quality level of services provided by the system. The more exacting the user requirements, the more complex is the necessary service system. This is reflected in the demands for its actions.

The control activity performed by an element or by a complex of elements in a communication system is called a *function* (sometimes accompanied by the adjective "communication"). For example, the function of protecting transmission against the influence of noise and dropouts is error control, or error detection and error recovery. The word function has been chosen because such an activity transforms states and/or values of performance measures into states and values being kept within prescribed limits. Moreover, some functions can be explicitly expressed by a mathematical formula, as for example more or less intricate relations between a data circuit error rate and a residual error rate gained by introducing an error control function.

Each function usually has some purpose assigned to it. The object of this purpose is achieved by means of operations which maintain the selected states and performance measures. Of course, not all performance measures are influenced by one function. For example, error control influences residual error rate (positively) and throughput (negatively) but cannot affect reliability (expressed, say, by availability: see Section 7.2).

In order to achieve the required goals an algorithm must be selected from theoretically known methods, and proved in practice. For some functions several tens of methods exist and the designer must discover the most suitable one under given conditions.

The next step is the implementation. First, the place or places of implementation must be determined. If the selected method is to be capable of coping with varying conditions (most of them should), information on the current state has to be acquired, collected, transferred from remote places, processed (if necessary) and finally employed. The implementation must also take into account the starting points provided by a criterion based upon status information for decision making in the initiation of an action. Returning again to error control, this function is implemented in all communication stations using information extracted from received coded data and released after an error is detected.

Last, but not least, it remains to decide about implementation tools (hardware, software, firmware) according to the cost– performance trade-off.

Since the function characteristics are important and will be used subsequently the following summary is given:

- Objective – the result of the function
- Method – the means of reaching the objective
- Place of the function's action
- Materialization and implementation

In computer/communication systems and computer networks processing functions are distinguished from communication functions. In communication systems, including PDNs and ISDNs, a finer classification is recognized and therefore the adjective will be omitted.

The list of functions is not constant and is contributed to by authors, groups of specialists, designers and documents. Their enumeration is further impeded by subdivision of functions into components. For example, error control is divided into error detection, error notification, error recovery, numbering and acknowledgement. The following list is compiled from various authors and is not exhaustive:

- Error control (error detection, recovery, notification etc.)
- Flow control (node-to-node, end-to-end)
- Routing
- Addressing
- Connection establishment and release
- Multiplexing and demultiplexing
- Splitting and recombining
- Formatting (framing and reassembling, blocking and deblocking, concentration and separation)
- Exceptional state recovery

An overview of classification schemes should follow. However, such schemes are dependent on time and authors, and do not help to solve the complex problems of communication control. The only established distinction with respect to the layered architecture is that between communication functions and transmission functions, but such a classification will rather be applied to services.

Results of function actions presented to users and seen from outside the communication system are called *services*. A service is commonly understood to be a gratification of requirements qualified in advance, but here it should have a more precise definition. In general terms a service is a certain capability of a communication system or network, which can be measured, evaluated and compared. As there is not a unique function, there is not a unique service, but a list of services; for example:

- Connection/connectionless-mode
- Normal/expedited data transfer
- Addresses
- Sequencing
- Confirmation
- Exception reporting
- Values of performance (quality of service) measures

Theoretically it is possible to assign each service a function or functions that support the service. For example, a connection is supported by connection establishment/release, routing and addressing, while data transfer may need multiplexing, flow and error control, formatting, and so on. Two classes of services are recognized:

- *Transmission services*, which provide for transparent (application-independent) and reliable data transmission between two or several users with a certain explicitly guaranteed quality (termed bearer services in the ISDN environment)
- *Communication services*, which add to transmission services a certain amount of data processing, transformation and storing for the benefit of users (ISDN teleservices)

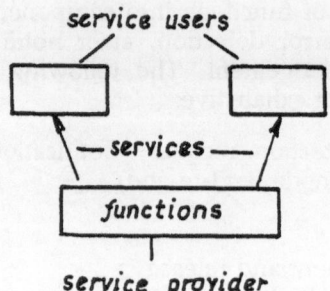

Fig. 4.1. The relationship between functions and services.

It must be admitted that transmission services may also include data processing and storage, but these operations remain within the communication system and are neither required by nor visible to the users. Note that the two classes of service refer to the well-known distinction between data transmission and data communication.

The simplest model of functions and services is shown in Fig. 4.1. The communication system is defined with a service provider while users are service users. Functions are performed by the service provider and their effects are perceived in the services.

The idea of transmission and communication services is often confused with the idea of telecommunication services. A telecommunication service is a set of capabilities provided by any telecommunication system (voice, text, data, image) or WAN. Telecommunication services include CCITT services which refer to public telecommunication services provided by postal, telegraph and telephone (PTT) administrations or by RPOAs.

The following sections indicate the qualities assigned to services.

4.2 Methods of Some Functions

This section describes the methods most often used in networks to perform error control, flow control, routing, addressing and control of recovery from exceptional states. These functions have been chosen not for their importance but rather for the diversity of their methods. The list is not exhaustive because of the large number of functions (although in this respect error control is the most important) and is made as short as possible. It relates particularly to methods which are referred to in the following chapters (and throughout the book), to avoid the need for further explanation. We also avoid performance comparisons, to avoid depreciating standardized methods that are widely and successfully employed.

The authors hope this survey might help in the understanding of standardization documents and recommendations, where more detailed comment is often omitted.

4.2.1 *Error Control* [7,21,33,45,47,58,76]

Error control methods have been in use for many years in response to the demand for accurate reception regardless of the transmission medium used. The results of Shannon's classical work [67] were very quickly enlarged and adapted for use in practice, and the concurrent development of coding theory supported these methods. It is estimated that more than 55 error control methods and their modifications have been proposed, described and at least modelled [49]. Contemporary standards and their embodiment in corresponding products, however, utilize only some of these methods, described below.

As the name implies, error control aims at decreasing the error rate within allowed deterioration of throughput, delay and costs. Error control methods are implemented at each communicating station, on transmission as well as reception, and need only the information about error occurrence in the transmitter–circuit– receiver chain. The function is launched at the instant of error detection (see Section 4.1).

First, let us refer to error detection and the error correcting codes which form the basis of almost all methods. Among several hundreds of possible codes the linear (even parity) codes, nonlinear (odd parity) codes, block codes and recurrent codes can be used, although standards and practice prefer iterative and polynomial codes for error detection.

Iteration means that two or more codes are applied to data fragments in turn. The simplest iterative code arises by double application of the parity check code (vertical and longitudinal) and is used to advantage in the case of character-oriented user data. If data characters are equipped with a parity bit it is used directly for vertical redundancy checking and hence a single longitudinal parity character is created (the terms being taken from punched tape techniques).

Although iterative codes are efficient enough, polynomial codes are preferred at present, in particular when user data is bit-oriented. Polynomial codes are, as their name suggests, generated by a polynomial. Polynomial codes include the well-known class of cyclic and pseudocyclic (shortened cyclic) codes. A typical example of a generating polynomial is $x^{16} + x^{12} + x^5 + 1$ which appears in most standards and gives rise to the polynomial code.

The simple detection of errors is not sufficient for automatic communication systems, and error control (as opposed to error notification) must involve a mechanism by which detected errors are corrected. If two independent channels between stations are available, feedback error control methods are provided. If not (as in radio communication), or if the propagation delay is large (as in satellite links) the *forward error control* (FEC) method has to be used. FEC methods need more redundant error-correcting codes and more complex decoding tools. They are costly and consume much computational and storage space. Feedback methods are among the oldest adaptive methods and are given preference over forward methods whenever possible.

Feedback methods may be classified as follows:

- Decision feedback – based upon decisions at the receiving side where the retransmission of the data unit is requested whenever errors are detected after decoding. It is also called *automatic repeat request* (ARQ). (An interesting, but not surprising, fact is that ARQ was first designed, although not named, by Van Duuren in 1943)

- Information feedback – requiring the activity of the sending station which decides about retransmissions upon an agreed criterion. For example, all transmitted data is compared with all received data coming back from the receiving station over the backward channel (echo check), or only the check bits used at both the sending and the receiving sides are compared (loop check)

Many modified combinations of the two methods have been described and verified in practice (hybrid methods, memory ARQ), but decision feedback ARQ prevails.

Briefly, the three types of ARQ methods that play a leading role in data networks assume that data is conveyed in data units over the forward channel only, and two commands (ACC for acceptance and proceeding, REJ for rejection and retransmission) are transferred, self-contained, over the backward channel. The three types are shown in Fig. 4.2. The first method, known as "stop-and-wait", is the simplest one (Fig. 4.2a). The sending station sends one data unit and awaits a command from the receiving station. The channels evidently operate in the half-duplex mode.

If the full-duplex mode is available, the stop-and-wait method is replaced by the more efficient continuous or "go-back-N" method (see Fig. 4.2b, where $N = 2$). The sending station transmits data unit by data unit until it gets a command from the receiving station. This reactive command however, reaches the sending station later, when N consecutive data units have already been sent. If the command is REJ the sending station must go back to the data unit sequence by N units in order to retransmit the incorrectly received one. Since the receiving station has discarded all data units after the incorrectly received one, the sending station is forced to retransmit all data units starting with the data unit demanded. The number N depends on the propagation delay of the forward and backward channel and is determined before the communication is opened. To realize this principle data units have to be numbered (see below).

Further improvements are achieved by retransmission of only those data units that are demanded, rather than a sequence of units. The receiving station, however, must capture all data units with undetected errors and insert the retransmitted one in the right place within the sequence of received units. This method is often called selective repeat (Fig. 4.2c).

Pure ARQ methods may be further combined with numberings and acknowledgements, which results in a number of different error control methods.

To prevent losses and duplication of data units and to simplify recognition of those that have to be retransmitted and then inserted among the received ones, the numbering of data units is utilized, as mentioned above. It would obviously be ideal to number all units serially but because of limited space in the data unit format and variable user data volumes (number of units carrying user data) a method of cyclic numbering is preferred. Cyclic numbering means numbering modulo an integer M, so that data units are numbered in succession $0, 1, 2, ..., M - 2, M - 1, 0, 1, ...$. The necessary space for carrying the number in the data unit must be greater than $\log_2 M$. If M is chosen appropriately (for example, $M = 2^k$) this space will be fully utilized.

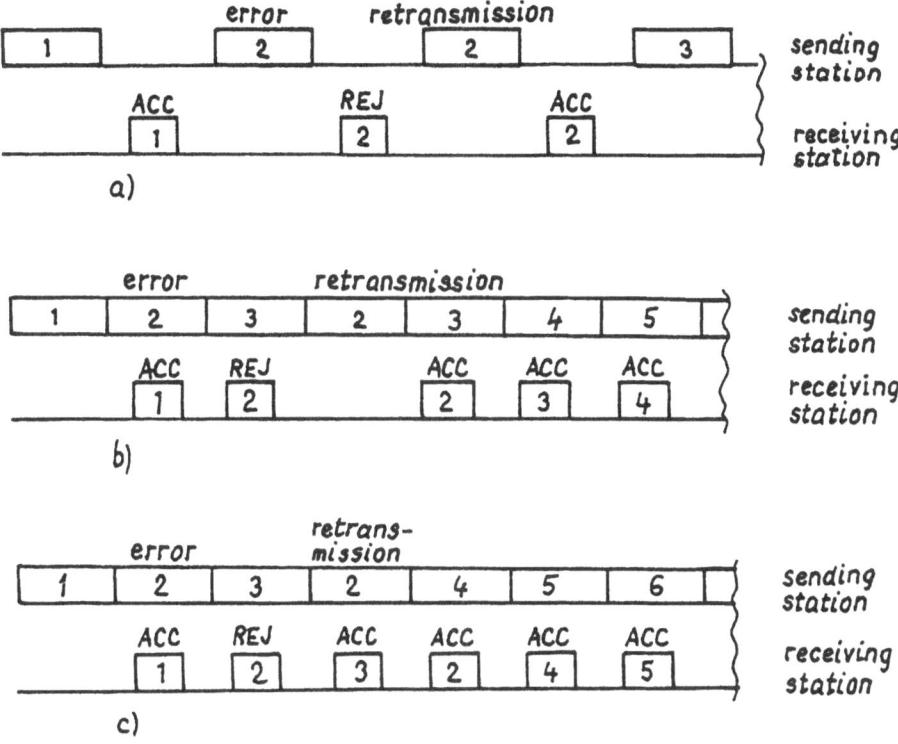

Fig. 4.2. Basic types of ARQ control methods. **a** Stop and wait; **b** continuous (go-back-2); **c** selective repeat.

The numbers denote either data units only or data units, commands and responses. The latter simplifies the continuous, as well as the selective, repeat ARQ methods (see Fig. 4.2b and c).

Up to now the transmission of only two commands over the backward channel has been considered. In order to avoid losses of units (data and command), a time interval is added. The time interval, called time-out (see Section 4.2.5), is agreed in advance by both stations and is timed by a timer started by the sending station at the end of each data unit transmission. If time-out elapses without any response from the receiving side, the sending station proceeds to a recovery (it retransmits the data unit). The time-out, however, can serve as one of the two commands: if it replaces ACC, the sending station retransmits only when it gets REJ (otherwise the data unit is considered to be accepted); if it replaces REJ the retransmission takes place only when the time-out elapses without any response.

This results in three acknowledgement schemes (methods):

- A-scheme or all-acknowledgement scheme, employing the three reactions (ACC, REJ and the time-out T)
- P-scheme or positive acknowledgement, employing only ACC and T
- N-scheme or negative acknowledgement, employing only REJ and T

The reaction of the receiving station may be required for each data unit as shown above, or for a group of consecutive data units. In the latter case the sending station is allowed to transmit a number of data units up to the limit agreed between both stations. This is called the *window size*. The transmission of further data units is prohibited unless a command appears on the backward channel. Since the window method is also used for flow control it will be explained more thoroughly in the following section.

4.2.2 *Flow Control* [7,14,33,34,45,47,50,54,58,72]

Flow control in PDNs has two objectives:

- To improve performance, by preventing loss of data and decreasing costs through optimal buffer allocation
- To avoid PDN congestion

This function is implemented in each communicating entity: in DTEs for end-to-end control, in network nodes for node-to-node and network-to-end flow control. The criteria of starting the function cover both the internal states of stations and nodes, and the external behaviour of communication media. Flow control employs the values of certain performance measures.

Flow control methods can be subdivided into fixed (static) control methods and variable (dynamic) control methods. Another classification criterion distinguishes between single and multiple data units operations.

A typical dynamic flow control method is throttling, when a receiving station governs the data transfer rate of a sending station by a special command (called a *choke packet* in the CIGALE data network of the French CYCLADE computer network).

Two dynamic methods concern single data units only. They are based upon the rejection or permission principles, both assuming control activity of the receiving station. In the former case the receiving station rejects data units by negative acknowledgement commands until it is capable of receiving. During the rejection state all data units are discarded. The latter case is similar to rejection: the receiving station releases the rejection by replying with a permission command for a single data unit (wait-before-transmission). The sending station, however, may become more active and ask for permission to send a single data unit (ask-and-wait). The simplest static method is directly involved in the ARQ stop-and-wait error control method described above. From the point of view of flow control it is in fact a combination of rejection (REJ) and permission (ACC). The rejection and permission principles can be extended to multiple data units whose number is either constant (agreed beforehand) or variable (when its value must be contained in each permission). The former case, known as a *window*, was first utilized in the ARPA network; the latter case uses the so-called *credit* method.

The window method consists of the following: a fixed number of consecutive data units numbered modulo M form a window. To avoid the appearance of data units with identical numbers within the window (the fulfilment of this requirement improves the ARQ selective repeat method of error control) the window size W – the figure representing the number of data units within the window – must not exceed M, the modulus of data units numbering. The window size can also be expressed in terms of window edges: if the left and right window edges are denoted by the lower number M_L and upper number M_U, respectively (Fig. 4.3a), then:

$$W = M + M_U - M_L + 1 \qquad \text{mod } M \text{ for } W < M$$

All data units within the window have permission to be transmitted (Fig. 4.3b, where $M = 8$ and $W = 2$). After exhausting the window size the sending station must wait for new permission and hence the data transfer rate is slowed down (see the delay between data units 1 and 2 in Fig. 4.3b). If the receiving station interrupts reception (for example, in response to error detection or overload) the window is set up in a new position (the left edge is shifted back to the last data unit successfully transmitted) but the window size remains unchanged (data unit 3 in Fig. 4.3b).

The window size is agreed for either a contractual period or for an individual communication. However, during communication it cannot be adapted to the actual situation in the network. This shortcoming is eliminated by the use of credits. A credit is a permission which is dispatched by the receiving station and contains an indication of the number of data units allowed to be sent by the opposite station. This number can be compared to a window of variable size.

Note that most flow control methods can be combined with error control (or vice versa).

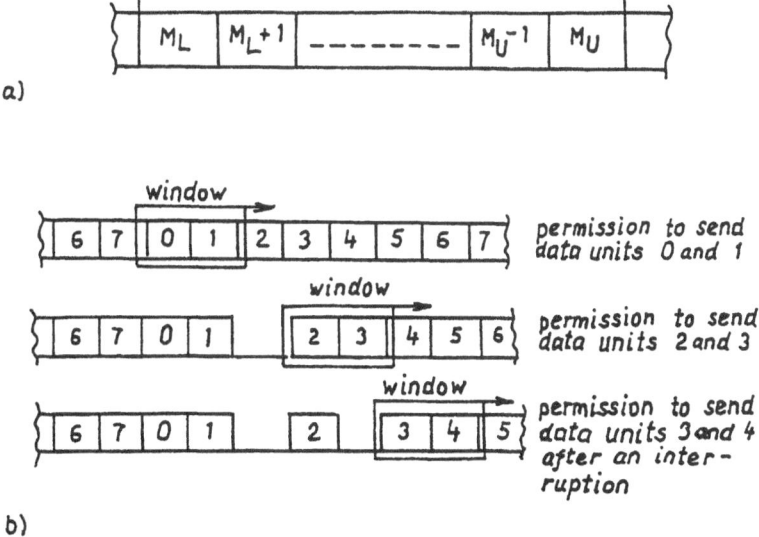

Fig. 4.3. The concept of the windowing mechanism. **a** Window size W; **b** an example with $M = 8$ and $W = 2$.

4.2.3 *Routing Control* [7,14,19,45,47,53,54,58,65,72]

The term *routing control* generally implies the control of the search for, and best choice of, an output route (direction) from a switching node following a demand to create a path (physical or virtual) for one or several data units to reach the required destination. This approach is independent of switching type: both circuit switching and packet switching need a control mechanism in each switching node in order to optimize the allocation of demands to communicate and the sharing of common resources throughout the network. Routing control usually involves the assignment – to each data unit entering a node – of an output line leading to the destination. Such assignment is the essence of packet switching. Although some of the routing methods are employed for circuit switching as well, we reserve the term *routing* for packet switching and replace the general term *data units* by the more specific term *packets*.

Routing control aims to optimize performance within prescribed limits, with particular regard to decreasing average transit time and increasing reliability. Routing control is implemented in each switching node and concerns each packet with the exception of packets for which the node is a terminal node (in this case the packet is directly routed to its destination).

Certain information has to be stored in each node to facilitate routing control: information about the status of the node itself as well as about that of neighbouring nodes, and, in some cases, about all the nodes in the network, information about the status of the node's elements (queue lengths, free buffer capacities, delays), about lines (only outgoing from the node and entering neighbouring nodes or all over the network) and, of course, about addresses, which should be monitored periodically or driven by events. Routing control is therefore a function requiring as much knowledge about the network as possible.

There are several classification schemes of routing control methods. We subdivide these methods into three groups:

- Random
- Fixed
- Adaptive

The random method involves selecting directions according to a certain probabilistic law. All directions may be equally probable (uniform distribution), or the probability assignment may prefer some of them. Even flooding can be regarded as being random although the routing mechanism is simply fixed (packets for which the node is not terminal are broadcast to all directions except the one from which the packet arrived). However, flooding leads to paths being randomly formed to the addressee. Random methods do not need any information about network state (the packet addresses are sufficient) and in any case they are resistant to failures in the network. But they neither guarantee network transit time nor conform to traffic changes. Nevertheless, some random methods have been introduced since 1964 in military data networks.

Both fixed and adaptive routing methods require the existence of routing tables in switching nodes. A routing table consists of a vector to all destinations to which the directions (outgoing lines, adjacent nodes) are

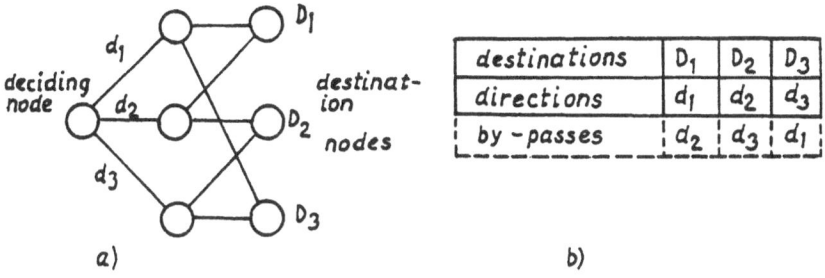

destinations	D_1	D_2	D_3
directions	d_1	d_2	d_3
by-passes	d_2	d_3	d_1

a)   b)

Fig. 4.4. An example of a routing table. **a** Part of a network; **b** routing table in deciding nodes.

assigned (Fig. 4.4). If the routing table is evaluated in advance from the static network structure (topology, transit delays, buffer capacities) and remains unchanged, the routing is fixed. Such routing methods are vulnerable to line and node failures, but can be modified by the addition of bypasses (alternative routes), at least one for each direction. In other words single path routing is replaced by bifurcated routing and the routing table increases by an additional row (in Fig. 4.4b, dotted lines). The choice between two alternatives may be made at random (random routing) or governed by the actual state of the switching node environment, for example, the size of output queues in assigned directions. This size is monitored and on reaching a certain critical value it invokes a switch-over (overflow routing).

The fixed method already contains features of adaptive routing, which is the most effective, but needs a mechanism for calculating entries of routing tables from information gained. Of course, adaptive methods are much more complex and the overhead (processing time, memory space, line capacity) is higher than in the case of non-adaptive routing.

The various adaptive routing methods differ in the way in which the routing table entries are derived, how often they are updated and who is responsible for the success of the routing process. For example, the well-known "hot potato" method utilizes the principle of the shortest queue, regardless of the destination direction. Routing in the ARPA network in its early stages of operation was based upon the values of the minimal delays between neighbouring nodes only.

The methods described above belong to distributed systems, where each node monitors its neighbourhood, updates its own routing table and eventually distributes the routing information to other nodes in the network. The routing information and routing table generators can be concentrated in a special node – *the network control centre* (NCC) which provides the network management functions – and the distributed routing becomes centralized. The NCC performs most of the activities of adaptive routing by virtue of status information gathered from switching nodes. It processes the incoming data, calculates routing tables and sends them to individual nodes as updates. Switching nodes are relieved of the routing decision activities and become as simple as in the case of fixed routing. They are required only to pick up the routing information and replace the obsolete routing tables by the updated versions.

Most PSPDNs utilize centralized routing methods even though they do not provide the updating quickly enough. To raise network performance, routing activities are distributed among switching nodes which, because of modern technology, does not increase costs substantially.

4.2.4 *Addressing* [11,21,38,45,48,54,58]

Naming and addressing are among the most crucial and complicated issues in networking. Unambiguous identification not only of each subscriber but also of many network components, elements, entities, points of attachment, provision of directory services, and processing (encoding, transforming) name and address representations whatever route the data will take over several interworking networks, are all involved in this function. "Name" as a linguistic notion is a construct identifying any object. It should be unambiguous, in that it identifies only one object. Unambiguity of a name, however, does not preclude the existence of synonyms.

Usually, "address" describes the location where a certain object can be found. In order to permit a clear identification within communication systems and networks, the address also has to be unambiguous throughout the whole network or the interconnection of several networks.

In the public network context a network address refers to the point of attachment either between subscriber and network at the DTE–DCE interface (DTE address) or between networks, or to the point at which the network service is offered. In Section 4.3 another term – network service access point – will be explained. This is another context within which the network address is considered.

There are three interpretations concerning the address representation:

- Abstract syntax
- External syntax which is involved in humanly-readable directories (so-called titles)
- Address encoding used to convey addresses during communication (signals in circuit switched networks, codes in packet switched networks)

A set of all addresses forms a global network addressing domain. The uniqueness of addresses in a domain is guaranteed by an authority associated with the domain (organization, administration, agreed document, etc.). The authority assigns and administers all addresses in one or several domains. The responsibility of authorities covers only associated domains.

Two different types of addresses are known that influence the address domain partitioning and address structure. The so-called flat address is not related to the location because the address cannot be abbreviated within the frame of, say, one country, and is not useful within an interworking environment but has been applied in some local environments (for example, the LAN known as Ethernet).

The hierarchical structure bears upon the global addressing domain partitioning and ensures a close relationship between the address and its location. The global network addressing domain is divided into disjoint subsets – network addressing domains. The network addressing domains may be subdivided into network addressing subdomains, etc. (see Fig. 4.5).

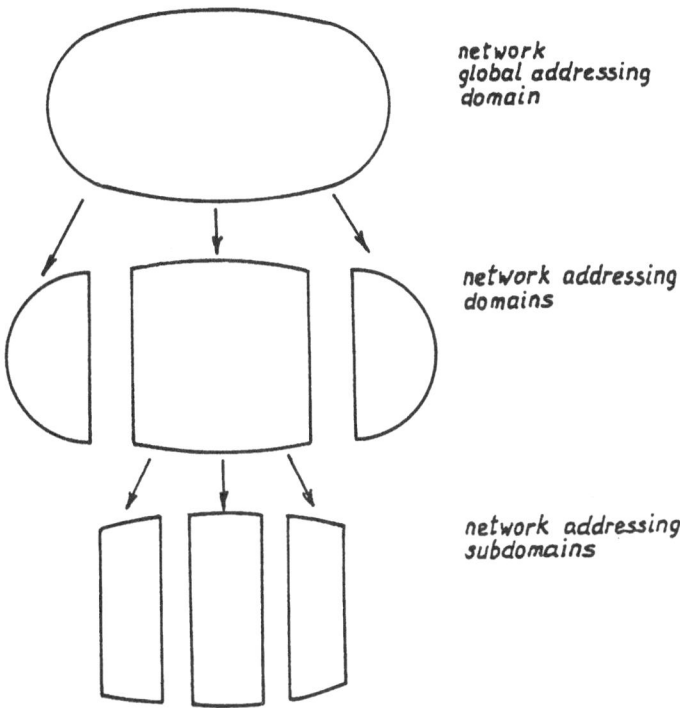

network
global addressing
domain

network addressing
domains

network addressing
subdomains

Fig. 4.5. Hierarchical structure of addressing domains.

Each addressing domain is administered by an authority which may register addresses itself or delegate subdomains (if any) to other subauthorities. However, the authority must ensure that addresses registered by sub-authorities are unambiguous. There are several ways of guaranteeing this. One way is to allocate a subset of addresses to the total set of addresses controlled by the authority. Another is to define an address component to be added to the address determined by the subauthority. The latter is applied in public network numbering plans (Telex, PSTN, PDN and ISDN), which are just the network addressing domains within the frame of the global network addressing domain.

In the case of hierarchical structuring the authorities of domains at the same level operate independently, but subject to the common rules imposed by the superior authority of the domain at an adjacent higher level. The conceptual structure of addresses may also follow the principle of hierarchy: the initial part of the address unambiguously identifies the domain, the next part identifies the subdomain (if any) and the rest is allocated by the subauthority. This may not imply, however, the explicit separation of address structure by means of special separators.

As an example consider the following address structure. The initial domain part (IDP) is a network addressing domain identifier specifying a subdomain and identifying the subauthority which may further structure the remaining part (the domain specific part, DSP). The IDP specifies its format and network addressing authority (authority and format identifier, AFI) and identifies the network addressing domain for which values of the DSP are allocated, and the corresponding subauthority (initial domain identifier, IDI). The address structure is treated in detail with respect to the PDN and ISDN numbering plans in Section 6.6. Addresses and their structures may or may not take into consideration the routing control, so routing information may or may not carry information about services, including the values of performance measures provided by a network.

The addresses do not often remain fixed during the communication. They are processed and transformed, in particular if the communication is routed over several networks as well as through modular-oriented systems. Transformations such as one-to-one, one-to-many, many-to-one, and many-to-many mappings used to be necessary for multiplexing and splitting functions. The corresponding methods are, however, beyond the scope of this book and will be mentioned briefly, where necessary, in the following chapters.

4.2.5 *Exceptional State Recovery Control* [58]

Communication control, however sophisticated, is not able to cope with all situations that may occur. It is to be noted that the network is not a deterministic system. Its behaviour is to a great extent influenced by noise, breakdowns, failures and varying data traffic, all more or less random in character, often following unknown probabilistic laws. Theoretically it is possible to introduce appropriate control tools to handle nearly all (but never all) situations. However, the control structure would be complex and would scarcely be feasible.

Most events which occur or may occur during a communication are covered by control by instruments. Such events are transmission errors and traffic overflows. An exceptional state is a state which is not foreseen because of its infrequent occurrence or simply because of its absurdity. Human communication controlled by intelligent beings resists the consequences of such events because a man or woman is able to find a way out of the exceptional situation. In computer communication, when human presence is suspended as much as possible, all actions have to be prepared in advance, otherwise the communication will be locked up whenever an exceptional state arises.

Control of recovery from exceptional states solves two problems:

- Recognition of the exceptional state
- Passage from it to another state, well defined and under control

The most convenient method of identifying an exceptional state seems to be the detection of threshold values of selected performance measures being exceeded. Nevertheless, more drastic mechanisms are used in practice with the aim of simplifying control and accelerating a counteraction. Since communication control is based upon network station handshaking, two identification mechanisms are very clear:

- The number of consecutive repetitions of activities (repeated sending of an identical data unit, command)
- A time elapse (time-out) from emitting a data unit or command to the reception of a response

Well-judged values of these two parameters lead to correct identification, although they are time-consuming and decrease efficiency. Nevertheless they are preferred for their simplicity.

On detection of an exceptional state, two follow-up procedures are available:

- Exceptional state notification (presented to a person, to a higher component of control)
- Passing to another well-controlled state (initial, quiet, ready, terminal, etc.)

The only problem is avoiding lock-ups and loss of user data. These demands often make communication control complex and costly.

4.3 Principles of the Layered Architecture [12,23,26, 27,31,37,45,47,51,52,59,72,74,79]

As has been emphasized, the complexity of communication control calls for partitioning. This complexity is one very important reason, but there are others. These are brought about by the variety of hardware, resulting in the use of many different manufacturers and software engineers, in the endeavour to obtain flexibility and interchangeability in response to increasing user demands and new techniques and technologies. Then there is the standardization activity.

In 1987, ISO technical committee TC 97 (Information Processing) recognized the call for standards on networks of heterogeneous systems and set up the subcommittee SC 16 (Open System Interconnection, OSI). The title was chosen to emphasize the networking (interconnection) and the opening to all systems worldwide obeying the same standards.

The SC 16 inaugural meeting was held in Munich in March 1978 and it decided to give the highest priority to the development of a standard (reference) model which would constitute the framework for the works of a protocol hierarchy. After 18 months of discussions and work the task was completed and the draft OSI reference model was passed to ISO TC 97. Finally, after the mandatory ISO procedures, the international standard ISO 7498 was approved in October 1984.

The OSI reference model was also recognized by a Rapporteur's Group of CCITT SG. VII on public data networks. The first draft of the OSI reference model for CCITT applications differed substantially from ISO 7498 but after certain modifications during the study period 1980–1984, it was approved by the VIIth Plenary Assembly in October 1984 as recommendation X.200. This recommendation is in fact a re-issue of ISO 7498 except for the title and some remarks.

This chapter cannot quote the documents cited above in full, but describes the main ideas of layered architecture, particularly with respect to the rest of the book.

Apart from functions and services there are three basic elements of OSI: entities, systems and layers. An *entity* is an indivisible element which is active

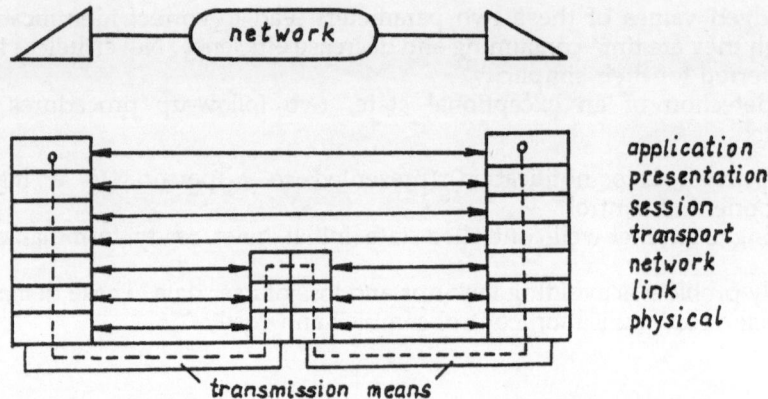

Fig. 4.6. Seven-layer model of the interconnection of two DTEs via a one-node network.

in performing a function and provides, alone or in co-operation with other entities, a service and resides in systems. Representations of computers or complexes of computers, the associated software, peripherals, terminals, human operators, physical processes, transmission means, etc. are considered as *open systems* within OSI. Interconnection of systems and networks with users forms the well-known and, in the past commonly used, horizontal partitioning. This has now been superseded by vertical partitioning. Each system is divided vertically into subsystems made up of one or more entities which interact directly only with entities in adjacent higher and lower subsystems of the same system. All subsystems of the same rank form a *layer*.

Fig. 4.6 shows the two methods of partitioning. Real systems, such as users with DTEs and a network, are represented as three systems: two end systems (DTEs) and one intermediate system (the network as a relay). The vertical partitioning into seven layers is the OSI reference model.

The basic principles of layering are:

- Entities in a layer perform a set of functions using services provided by entities of only the adjacent lower layer (with the exception of the lowest layer)
- Entities in a layer provide services to entities in the adjacent higher layer (with the exception of the highest layer) independently of how these services are performed
- Each layer may be subdivided into sublayers with equal properties – sublayers can, if necessary, be bypassed with the exception of bypassing all sublayers of a layer
- Co-operation between entities in the same layer but of different systems (so-called peer entities) is governed by one or more layer *protocols* (peer protocols)
- Co-operation between entities of adjacent layers within a given system is provided by means of *service primitives* through logical interfaces called *service access points*, identified by addresses

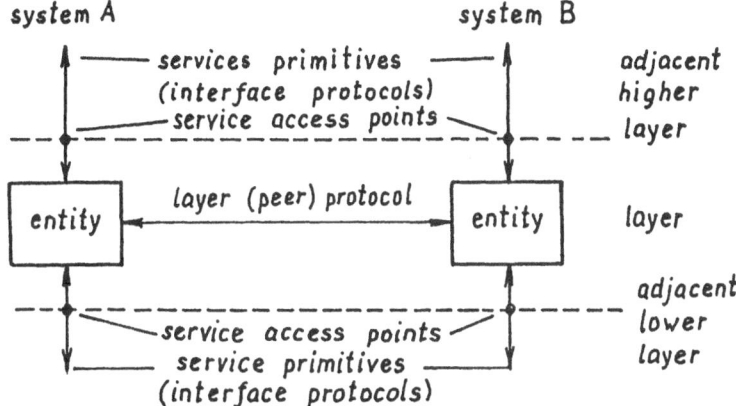

Fig. 4.7. The relationship between basic terms of layering.

Fig. 4.7 shows the relationship between layers, services, service primitives, protocols, etc. Some of the ideas presented in Fig. 4.7 require at least a brief explanation. Within a given system each service is provided by an entity acting at a service access point. Each entity performs a function or functions by relying on services provided by the adjacent lower layer and by communicating with peer entities located in other systems. Thus each layer is able to add some value to services provided by lower layers and supported by all systems so that the highest layer is offered the full set of services with the necessary quality to run distributed applications.

Comparing Fig. 4.7 with Fig. 4.1, we may consider each layer relatively as a service provider for the layer above, which is then a service user. Hence, each layer provides a certain set of layer services.

There are three categories of layer services (compare with network services, see Section 5.1):

- Mandatory service (which must be provided)
- Provider optional service (which may or may not be provided)
- User optional service (which is provided only if the user requests it)

Each service user interacts with the service provider by issuing and receiving abstract, implementation-independent, primitives. Basic primitives are request, indication, response and confirmation. Each of these has a corresponding direction and being associated with values of selected parameters. Types of primitives, associated parameters and time sequence diagrams of events at service access points are topics of abstract layer service definitions (without any implementation specification).

There are two types of associations:

- Associations between peer entities governed by layer protocols
- Associations between entities in a subsystem and the next lower subsystem of the same system governed by interface protocols

The communication passes vertically from the highest layer of an end system to the physical medium (top down through all subsystems) and again from

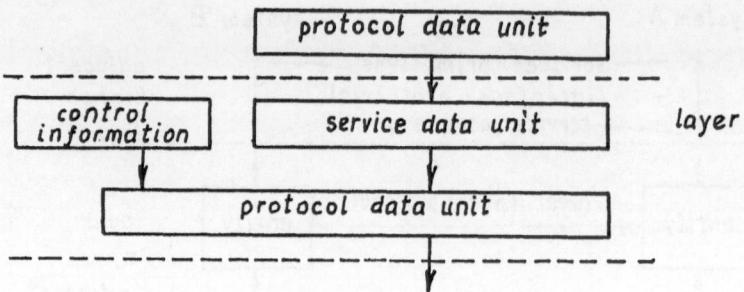

Fig. 4.8. The creation of protocol data units.

the medium up to the highest layer of the adjacent intermediate system (bottom up through all its subsystems), as shown by the dotted lines in Fig. 4.6. To make such a communication possible, protocols are needed for communication within layers across all interconnected systems.

Both ways of communication need "wagons" to convey user data and the necessary control information. These means of transport are called *data units*: protocol data units for layer protocols and interface data units for interface protocols. Each formatted protocol data unit in a layer has to be mapped onto one or several service data units in the adjacent lower layer, forming new user data units. The control information for co-ordinating the joint operation between peer entities in the layer is added to the user data units. The complete formatted data unit now created gives rise to a new protocol data unit for a lower layer protocol (Fig. 4.8). These operations continue to increase protocol data units up to the lowest level or subsystem of the same system. In an adjacent system the roles are changing and the size of protocol data units again decreases so that in the highest subsystem of the opposite end system the original data units reach the user.

It follows that the protocol is a set of formats: that is, the arrangements of bits (bit-oriented format) or characters (character-oriented format) in the protocol data units, rules for the exchange of protocol data units between peer entities (reactions on events), and *parameters* – measurable variables whose values are set by agreement between entities before a communication (time-outs, number of repetitions, user data volumes, window size, numbering modulus). The protocol specifications result from linking up standardization documents.

The associations are established for data transfer between two (or more) entities and give rise to point-to-point (multipoint) connections. Connections between service access points are set up in a layer upon demand of entities of the next higher level before any communication in that layer can take place. Such a mode is called *connection-mode* and the communication always proceeds through the connection establishment (call set-up) phase, the data transfer phase and the connection release (call clearing) phase. The principal argument in favour of the connection-mode is that the necessary resources can be allocated, specified services and their quality can be agreed and guaranteed, the overheads associated with addressing can be avoided during the data transfer phase, and sequencing and flow control can be maintained. Connection-mode communication is attractive in applications that call for relatively long-lived and stream-oriented data transfer in stable conditions.

The counterpart of connection-mode is *connectionless-mode*, which is also known as "datagram" or "transaction mode". It involves the transmission of data in a single self-contained operation without establishing, maintaining and releasing a connection, over pre-existing associations where necessary resources are only dynamically assigned. This mode cannot guarantee the relationship (sequencing) between data units at the destination side and may not necessarily report their non-delivery.

This brief introduction to the principles of layered architecture permits us to construct models of data communication systems or networks. In order to avoid self-made constructions the seven-layered reference model was proposed, primarily to cover the distributed data communication system based on packet switching, although other applications followed shortly afterwards. The "magic number" seven was not chosen by mere chance. It was derived from the principles described in the corresponding standardization documents (ISO 7498 and CCITT X.200). The reference model, once established, may no doubt suffer from slight shortcomings, but it should be followed as much as possible, in particular because many standardization documents have already been attached to it and have become valid. Moreover, the world computer communication market offers so-called OSI products which strictly obey the rules of the reference model.

The OSI reference model is described in separate layers, starting from the top with the application layer down to the physical layer (Fig. 4.6). The layers are described below, with particular reference to data networks and CCITT services.

The *application layer* provides information exchange and remote operations required by application processes. It serves as a window between these processes and concerns the semantics of applications. Application entities perform functions necessary for exchanging semantically meaningful information, such as remote access to data files and databases, and remote transfer of texts, graphics and images. Many application layer protocols have been or are being developed: access and management, job transfer and manipulation, virtual terminal services, message handling services, telematic services such as teletex, videotex, etc. Application services differ from services provided by lower layers because of their top position. They are not provided by any other layer nor are associated with a service access point. The list of application services is not only long but also open for future applications and demands.

The primary purpose of the *presentation layer* is to provide an application process independent of differences in data representation by selecting a mutually agreed specific syntax. This syntax is used to represent structured data during the bit-transparent data exchange provided by the adjacent lower layer. The presentation context is specific either to a particular application such as teletex and videotex, or to a particular type of hardware such as a machine representation (machine codes and alphabets). Another service is the transformation of syntax concerned with code and character set conversions, with the modification of data layout and the adaptation of actions in the data structures.

The *session layer* provides the mechanism for establishing, organizing and synchronizing the dialogue interactions between application processes. It allows for two-way simultaneous and two-way alternate communication and the definition of special tokens for structuring exchanges. The list of session services provided to the presentation layer contains normal, expedited and quarantined data exchange and exceptional state reporting. In the connection-mode the one-to-one, one-to-many and many-to-one mappings of session connections onto transport connections are defined. Corresponding session protocol data units are sometimes called *messages*.

The purpose of the *transport layer* is to provide transparent data transfer between end systems and thus to relieve the upper layers of any concern with the details of achieving a reliable, cost-effective data transfer. In many cases the boundary between the transport layer and the network layer represents the traditional boundary between the transmission services provider (such as PTT) and users or, in other words, between the communication and transmission services. From this point of view the transport layer optimizes the use of transmission services and performs the functions needed to meet the quality of service required by the session layer. For example, if the accuracy of data delivery offered by the network layer is below the value required by the transport layer users, the transport layer must add error control. If the costs of network connections exceeds the budgets of transport layer users, the transport layer should provide multiplexing of several transport connections onto a single network connection so that the costs may be shared by several users. The transport protocol data units are now called *blocks*.

The *network layer* provides data transfer between end systems over a switched network possibly involving many data circuits and relay (intermediate) systems. Thus, it relieves its users of relaying and routing considerations and makes them independent of data transmission technologies and switching technologies used in different networks. The network layer also handles relaying and routing of data over concatenated networks, if necessary, while maintaining the quality of service requested by the transport layer. The variety of services provided by the network layer is large, though some of them are optional (provider optional or user optional). If the packet switching service is provided, the protocol data units conveying transport layer blocks are *packets*. In circuit-switched networks they are hidden in *signalling*.

The *data link layer* provides for reliable data transfer over physical data circuits. This includes the functional and procedural means of transferring protocol data units called *frames* (earlier "blocks") as a service to entities in the layers above, and detection and correction of errors which may occur in the layer below. In addition, it enables the network layer to control the interconnection of data circuits within the physical layer.

The purpose of the *physical layer* is to provide for data bits (in serial transmission) or data characters/octets (in parallel transmission) to be transferred via the physical media. This includes the functional and procedural standards required to activate, maintain and deactivate the data circuit, possibly provided by interlinking different physical media. These standards are determined by the transmission medium used and therefore a number of protocols have been and are still to be developed. The real physical

Table 4.1. Standardization activity in relation to the layered model of networks

Layer	CCITT recommendations	ISO standards
Network	I.340, I.450 (Q.930), I.451 (Q.931), I.452 (Q.932), I.461, I.462, I.463, I.465, I.540, I.550, Q.704, Q.705, Q.711, Q.712, Q.713, Q.714, Q.761, Q.762, Q.764, Q.930, Q.931, Q.932, S.19, S.20, V.25, V.25bis, V.110 (I.463), V.120, X.20, X.20bis, X.21, X.21bis, X.25, X.29, X.30, X.31, X.32, X.70, X.71, X.75, X.80, X.81, X.82, X.96, X.213, X.223, X.300, X.301, X.302, X.305, X.320, X.321, X.322, X.323, X.324, X.325, X.326, X.327	8208, 8348, 8473, 8648, 8878, 8880, 8882, 9068, 9542, 9574, 9577, 10028, 10029, 10030, 10177
Link	I.440, I.441, I.462, I.463, Q.702, Q.703, Q.920, Q.921, S.15, T.71, V.41, V.42, X.25, X.28, X.32, X.75, X.212	1155, 1177, 1745, 2111, 2628, 2629, 3309, 4335, 7478, 7776, 7809, 7826, 8471, 8802, 8867, 8882, 8885, 10038, 10039, 10171
Physical	I.340, I.411, I.412, I.430, I.431, I.460, I.461 (X.30), I.462 (X.31), I.463 (V.110), I.464 (V.120), I.465, S.16, V.10 (X.26), V.11 (X.27), V.13, V.14, V.21, V.22, V.22bis, V.23, V.24, V.25, V.25bis, V.26, V.26bis, V.26ter, V.27, V.27bis, V.27ter, V.28, V.29, V.31, V.31bis, V.32, V.33, V.35, V.36, V.37, V.54, V.110, V.120, V.230, X.20, X.20bis, X.21, X.21bis, X.22, X.24, X.25, X.26, X.27, X.32, X.50, X.50bis, X.51, X.51bis, X.52, X.58, X.61, X.70, X.71, X.75, X.80, X.81, X.82, X.150, X.211, X.221	2110, 2593, 4902, 4903, 7477, 7480, 8480, 8481, 8482, 8802, 8867, 8877, 9067, 9314, 9543, 9549, 10022

medium is interfaced to the physical layer by specifications of the mechanical connector and electrical (and optical) signals. However, these specifications are beyond the scope of OSI.

Table 4.1 arranges, at least roughly, the spate of CCITT and OSI standardization documents concerning the OSI reference model (fully or partially) in accordance with the three lower layers, which are the subject of the following sections.

Besides layer control, OSI also needs for its functioning an arbiter capable of recognizing the problems of initiating, terminating and monitoring activities, assisting in their harmonizations and solving exceptional situations. These functions are collectively considered as *management* within the OSI architecture. They involve the collection, processing and distribution of information about the status of resources, their changes, and the co-ordination of management processes.

OSI management also deals with diagnosing faults and their recovery, tariffs and charging, network reconfiguration, reporting statistics for planning and analysis of resource utilization, security control, etc. (see Section 3.4). There are three categories of OSI management activities – system management, layer management and layer protocol management – which reside either in systems or subsystems, or are concentrated in a special oversystem equipped with a management information base.

4.4 Layered Models of Networks [8,23,36,40,57,69,72]

The OSI reference model is a very general model for the interconnection of different computer communication systems in any layout through means of communication that include means of switching. The real network is, however, only a part of the model, and hence the network with its end systems – DTEs – has to be dealt with. The subnetwork in OSI terminology which models the real network itself is not able to handle communication processes autonomously. It needs end systems equipped with all the higher layers in order to provide the OSI network or transmission services at the boundary between the network layer and the transport layer. This model is repeated in Fig. 4.9, with two end systems and two intermediate systems representing a specific network with its subscribers. As the term "network" would be confusing within the OSI environment, we shall use the abbreviation PDN to mean generally any public network providing data services (PSTN, CSPDN, PSPDN, ISDN).

In cases when users (represented by DTEs) communicate over a PDN, the PDN participates in performing the transmission services. This participation is seen at the DTE–DCE interface, because the service provider (PTT, RPOA, etc.) delivers the service at "the end of wire" rather than at the network layer. In order to characterize this approach, we use the term *data transmission services* or, if a PDN is available, *PDN services*. Such a distinction is acceptable because of the natural boundary between the user's and provider's responsibilities. As a rule, the user is responsible for the establishment and maintenance of DTEs while the PDN provider offers data transmission or PDN services. The user is then either satisfied with them or may implement additional functions in higher layers of his or her DTE in order to enhance communication capability.

This approach also leads to separate protocols: PDN protocols and user protocols. The former are subdivided into access protocols, which govern user access to the PDN and should be well known to users, and intranetwork protocols, which are not visible to users and are the provider's own concern. The two types of protocol may differ because the intermediate systems are

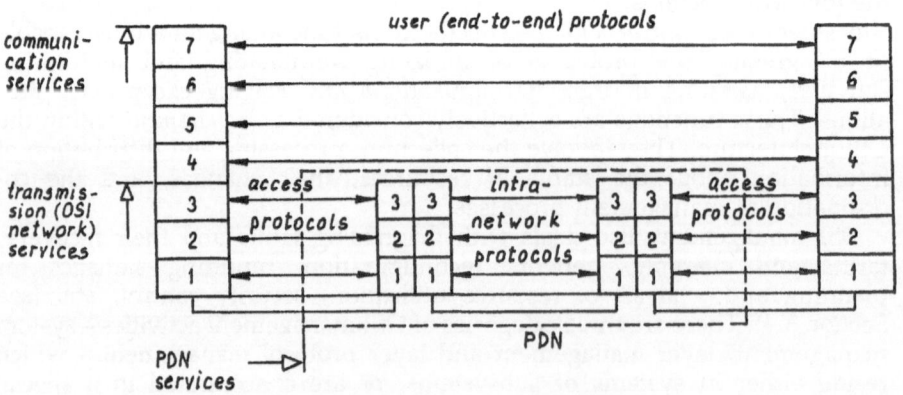

Fig. 4.9. The layered model of PDNs.

"Janus-faced" and may handle different protocols within their halves. User protocols constitute the well-known class of end-to-end protocols which are implemented by users even though they often keep to standards and commercially available products.

PDN modelling also projects, in terms of protocol, data units, which have already been mentioned in the previous section and are recapitulated in Fig. 4.10. Note the conventional (but here used in a specific application field) terms such as *fragment* (piece of user data which is conveyed over the PDN as a whole), *header* and, if need be, *trailer* (frame check sequence, FCS), all of which represent protocol control information.

Fig. 4.10 shows not only protocol data units with user data (the left-hand side of the figure) but also protocol data units with control data only for performing implemented functions (the right-hand side).

The OSI reference model was conceived to reflect aspects of packet switching. The modelling of other types of PDNs is a task which must respect on the one hand the properties of transmission and communication means that were exploited before the advent of OSI, and on the other the principles of the reference model which should be preserved. The most crucial problem has to do with the layering and, particularly, the attributes of layers.

In the OSI reference model one tacitly assumes that all layers act, to varying degrees, during the whole communication process. Some layers in fact perform functions and provide services all the time: the physical layer and the application layer are examples of this.

The data communication process, however, consists of several phases, especially in the connection-mode; that is, the call control and data transfer phases. Each phase requires some functions and services to be successfully executed. Since the functions within the reference model are spread among layers (some of them are even repeated in several layers) there may exist a phase which needs no function assigned to a particular layer. Such a layer could be called semi-active because implemented functions are not permanently performed.

In the layered model of PDNs yet another case occurs. During communication a layer within intermediate systems does not act at all, one that performs no functions and provides no services and seems to be transparent (horizontally) for end systems. This layer is therefore null or transparent. The latter term is more appropriate since it expresses the vertical transparency still needed because of the layering principle that no layer in the reference model may be bypassed. If, however, the primitives crossing the higher service boundary are mapped directly onto primitives crossing the lower service boundary (and vice versa), the services provided for the adjacent higher layer will be identical to the underlying services provided to the transparent layer. The layer transparency, however, influences higher level protocols which should take on functions necessary for the services required.

Another approach is to preserve all layers in intermediate systems from the highest active layer below and minimally use their functionality. This yields the possibility of reducing the number of protocols (just standard protocols are sufficient), but, as shown in Fig. 4.10, the overhead increases and the overall efficiency decreases. Nevertheless, this approach is preferred, so simplicity is sacrificed to flexibility and universality.

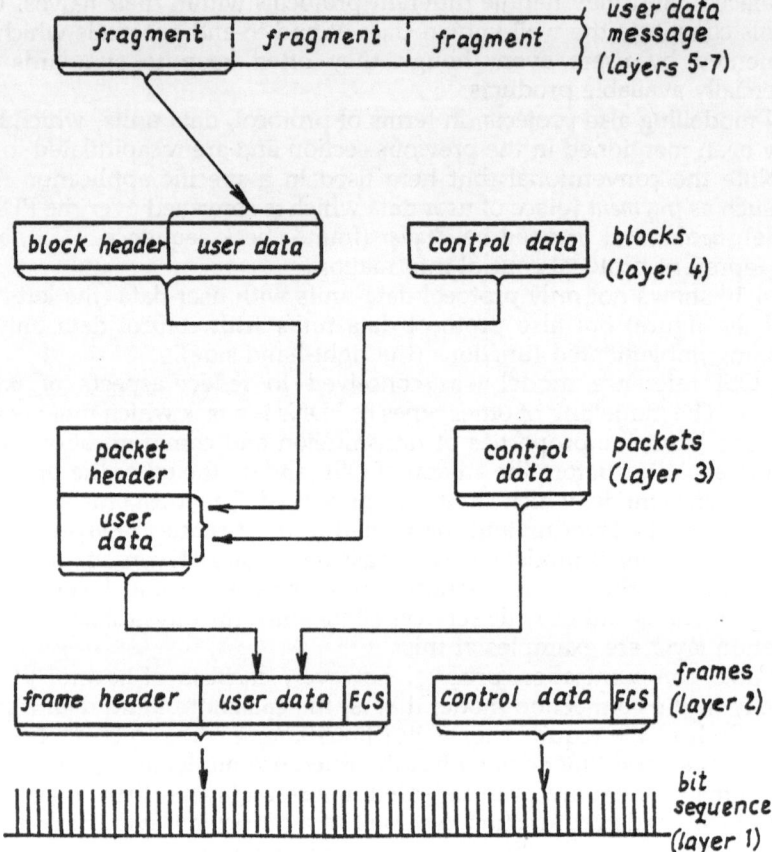

Fig. 4.10. Overview of protocol data units within a PSPDN.

Since CSPDNs appeared long before the layered architecture was developed, their layered models naturally face the problems mentioned above. During the call establishment and release phases only the network layer is active whereas during the data transfer phase the link layer becomes transparent and the CSPDN behaves like a leased circuit. The link layer needs an end-to-end protocol while the call procedure (establishment and the corresponding signalling) proceeds step-by-step through intermediate systems (switching nodes). Fortunately, the OSI link layer provides the function, conveying to the network layer the capability of controlling the interconnection of data circuits within the physical layer. Therefore this function, if implemented, facilitates direct control of a non-adjacent layer (in this case exceptionally between the network layer and the physical layer). The link layer, however, cannot be completely transparent (null).

The layered model of CSPDN with synchronous DTEs (SDTEs) is depicted in Fig. 4.11 with the aim of bringing the model closer to contemporary reality. The CSPDN layers are labelled by CCITT protocols (this convention will be applied in all models which follow). CCITT documents were chosen since only public networks are taken into account. The standard protocols will be expanded upon in the following sections.

Fig. 4.11 refers to the CSPDNs based upon analogue internetwork circuits and analogue or digital access circuits equipped with multiplexers, if any. For the link layer, standard protocols are recommended even though any end-to-end protocol could be used (ISO/BSC, CCITT V.41). The user protocols are not enumerated although they are approved.

The PSPDN models comply very well with the reference model in Fig. 4.9. A model under the same conditions is presented in Fig. 4.12, with the exception of DTEs which are replaced by packet-mode DTEs (PDTEs), and needs no comment.

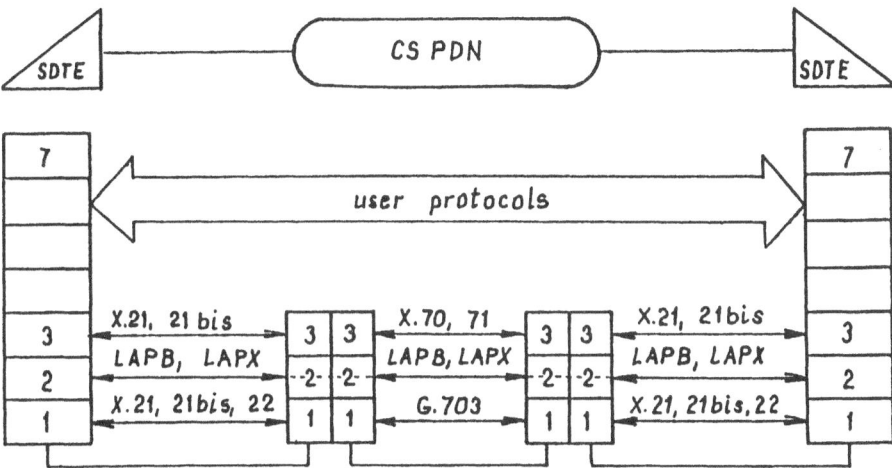

Fig. 4.11. The layered model of the CSPDN.

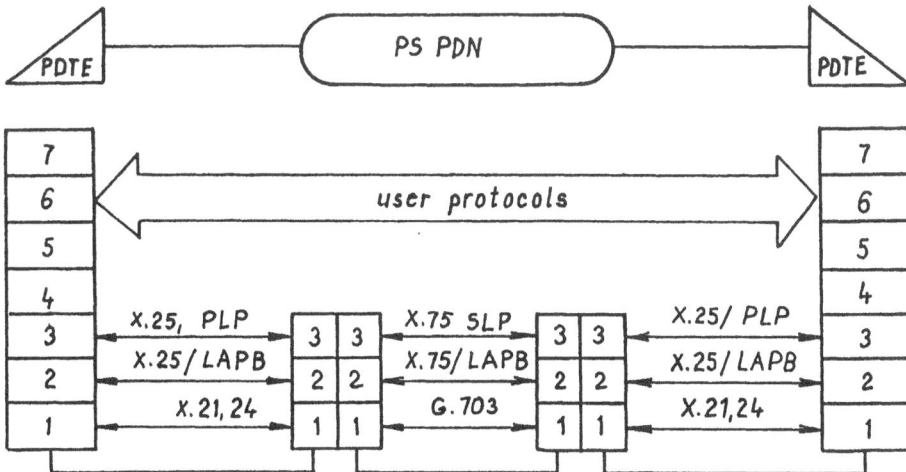

Fig. 4.12. The layered model of the PSPDN.

Fig. 4.13a and b shows possible ways of access to PDNs. Fig. 4.13a depicts the access of synchronous DTEs to the CSPDN through synchronous TDMs which perform only physical functions. Fig 4.13b is devoted to the analogue access of PDTEs to the PSPDN through modems and leased lines possessing the same functionalities as multiplexers. A very similar model holds for an access over the PSTN; only the corresponding protocols differ. For half-duplex transmission over the PSTN the LAPB protocol is extended as defined in CCITT X.32, and the network protocol involves the current telephone call procedure for automatic data communication as defined in CCITT V.25 or V.25bis.

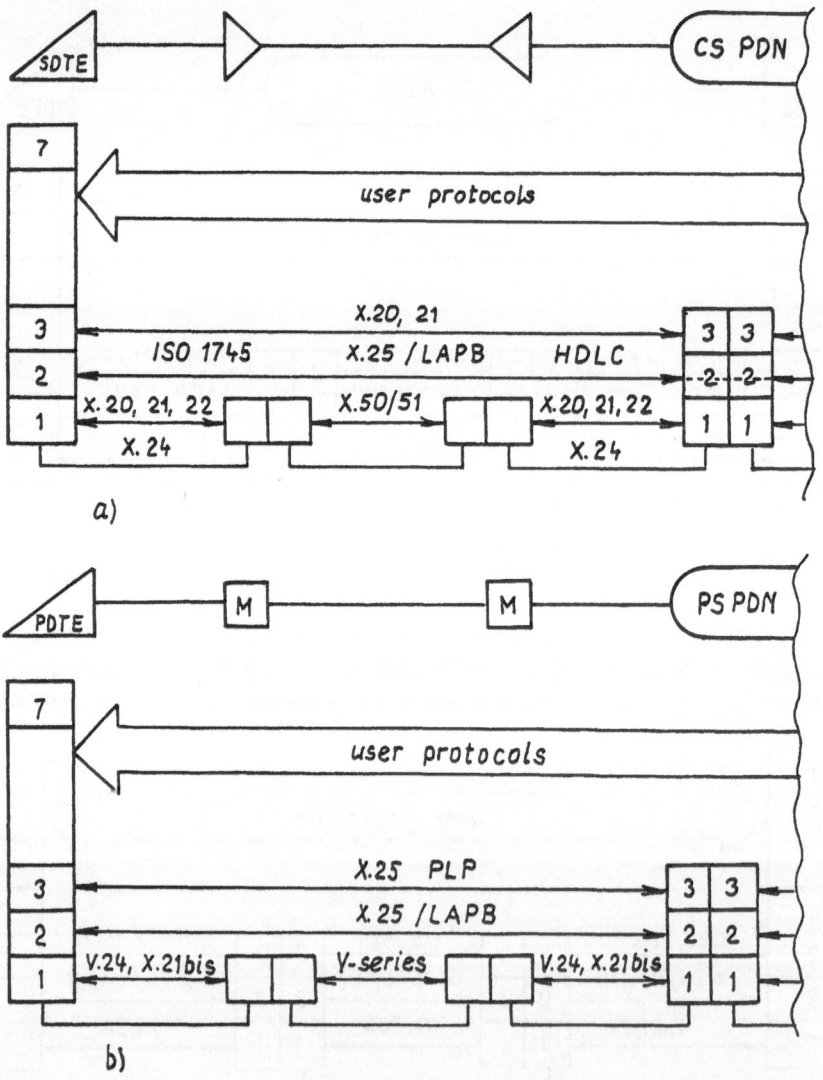

Fig. 4.13. Examples of access to PDNs. **a** Access to the CSPDN over a leased digital circuit through multiplexers; **b** access to the PSPDN over a leased analogue circuit through modems.

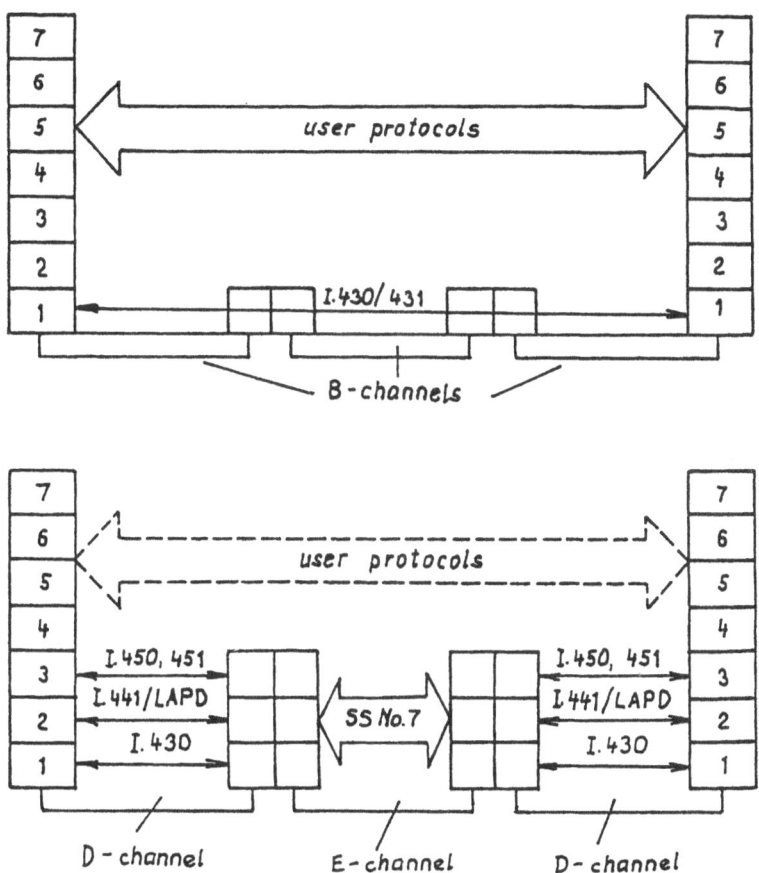

Fig. 4.14. The layered model of circuit switched data service on the B-channel within the ISDN.

The very familiar, and commonly utilized, access of start–stop DTEs (ADTEs) to the PSPDN is omitted because of the non-conformity of PADs with OSI. PADs are considered as non-OSI adaptors or devices because they were designed and elaborated earlier. A better designation would be "pre-OSI", but in spite of attempts to introduce them into the reference model, they remain outside. As explained in Chapter 6, two protocols have been introduced: CCITT X.28 for the character exchange between an ADTE and a PAD, and CCITT X.29 for the exchange of packets between a PAD and a remote PDTE or another PAD. Since both protocols partially deal with code conversions residing in the presentation layer and with session control, the corresponding model could hardly be adopted to the OSI environment. The only solution for the time being is to use the adjective "non-OSI".

ISDN modelling requires a slightly different approach stemming from the different structure and operations of ISDN (see Section 2.3). While PDNs have at their disposal a single communication path for the exchange of both user data and protocol control information throughout the whole communication process regardless of phases, the ISDN strictly separates the information (data) exchange plane from the control exchange plane. Thus, the two-dimensional representation is not sufficient. On the other hand, the ISDN was developed after the appearance of OSI and so it could adhere to the layered principles and profit by sophisticated protocols.

Fig. 4.14 is an example of two-dimensional projection of the ISDN model onto two planes. The B-channel handles user protocols beginning from the link layer, and the physical layer is transparent just after the call is set up and is governed by the CCITT I.430 or I.431 protocols for basic or primary access, respectively. The D-channel needs three layers for control and the remaining four layers may be employed by users for signalling and application purposes, if necessary (dashed lines in Fig. 4.14). The corresponding protocols are specified in CCITT I-series recommendations: I.430/I.431 for the physical layer access protocol, LAPD of I.441 for the link layer access protocol, and I.450 and I.451 for the connectionless-mode network layer protocol. The internetwork control is based upon signalling system No.7 (SS No.7) and the so-called E-channel, primarily developed for modern switching systems and digital circuits. The arrangement just described is suitable for the circuit-switched data service on the B-channel.

Models of two ISDN accesses for packet switched data services on either the B-channel or the D-channel are presented in Fig. 4.15. Moreover, the access on the B-channel requires a packet handler with a three-layer structure in order to handle packets. The access on the D-channel is very similar to the digital access to PSPDNs, and, together with the above model, needs no comment. Fig. 4.16 shows how the layered model is capable of reflecting the structure and equipments of the ISDN user interface. It points out three ways of access via interfaces S, T and U: single layer network termination NT2 (an active bus), double layer NT2 (representing, for example, a statistical multiplexer) and full (triple) layer NT2 (private branch exchange). Only the control exchange over the D-channel is shown; the data exchange plane does not differ from the layered model of the B-channel in Fig. 4.14.

Layered models are particularly effective when two or more networks are to provide data transmission or communication between remote users (DTEs). The variety of transmission supports which are organized not only internationally but also within one country calls for efficient and cost-effective interworking tools. Besides public means such as PSTNs, CSPDNs, PSPDNs and the emerging ISDN, there is a series of means demanded and even provided by users (mobile systems, satellite and land, private data networks). Each transmission support possesses specific properties which are offered to users as services and facilities of a certain quality. More or less independent development of networks gives rise to a variety of protocols and signalling systems, which results in a network service sometimes differing from the OSI network service.

Direct interconnection is not possible even between networks of the same type. The conversion of signalling and protocols including protocol data units is the minimum measure if the two networks have the same layered architecture. If services provided by a layer of one network are not sufficient and differ from services provided by the other network, an additional protocol which would concentrate both sets of services has to be implemented. The *interworking unit* (IWU) or *interworking function* (IWF) (both

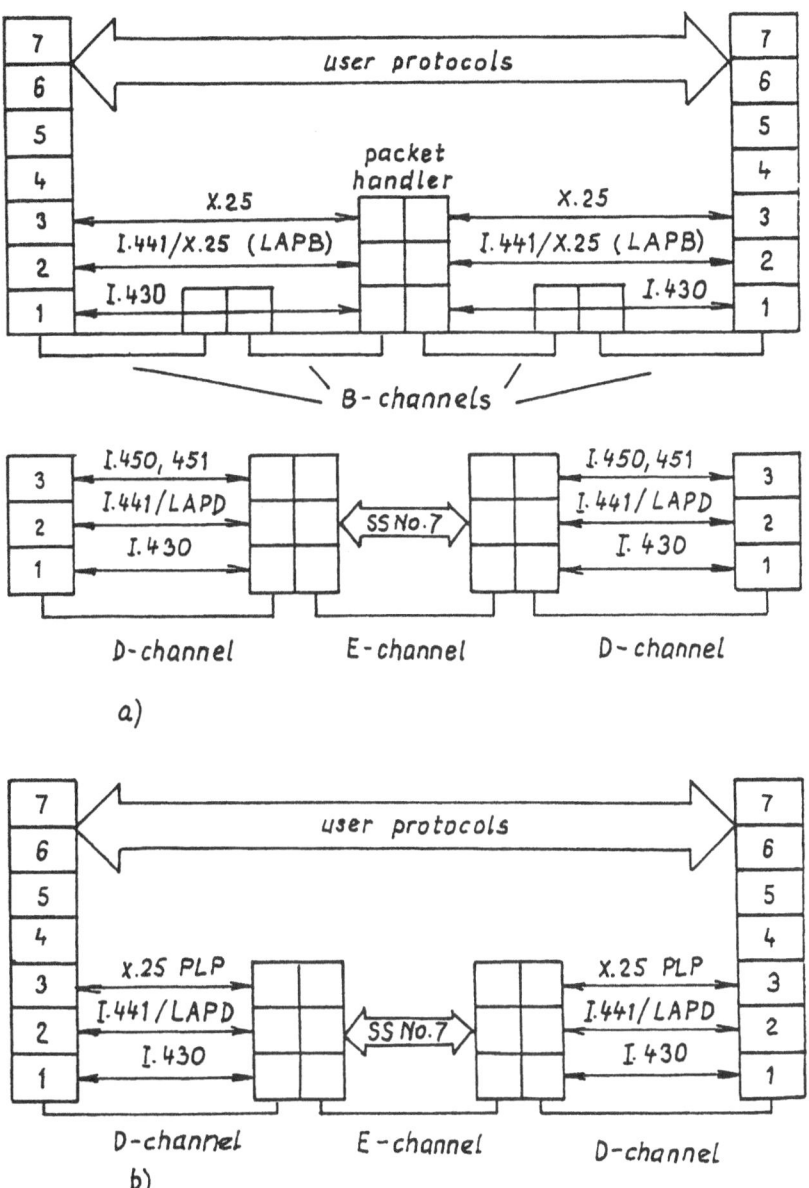

Fig. 4.15. Layered models of packet switched data services within the ISDN. **a** Access on the B-channel; **b** access on the D-channel.

terms appear in the literature) serves as a gateway or bridge at the interface of two networks and is provided by an access unit, PAD, terminal adaptor, and, in the simplest case, by the buffered modem. The interworking unit can be operated from both networks, be a part of one of them, or be functionally divided between the two networks.

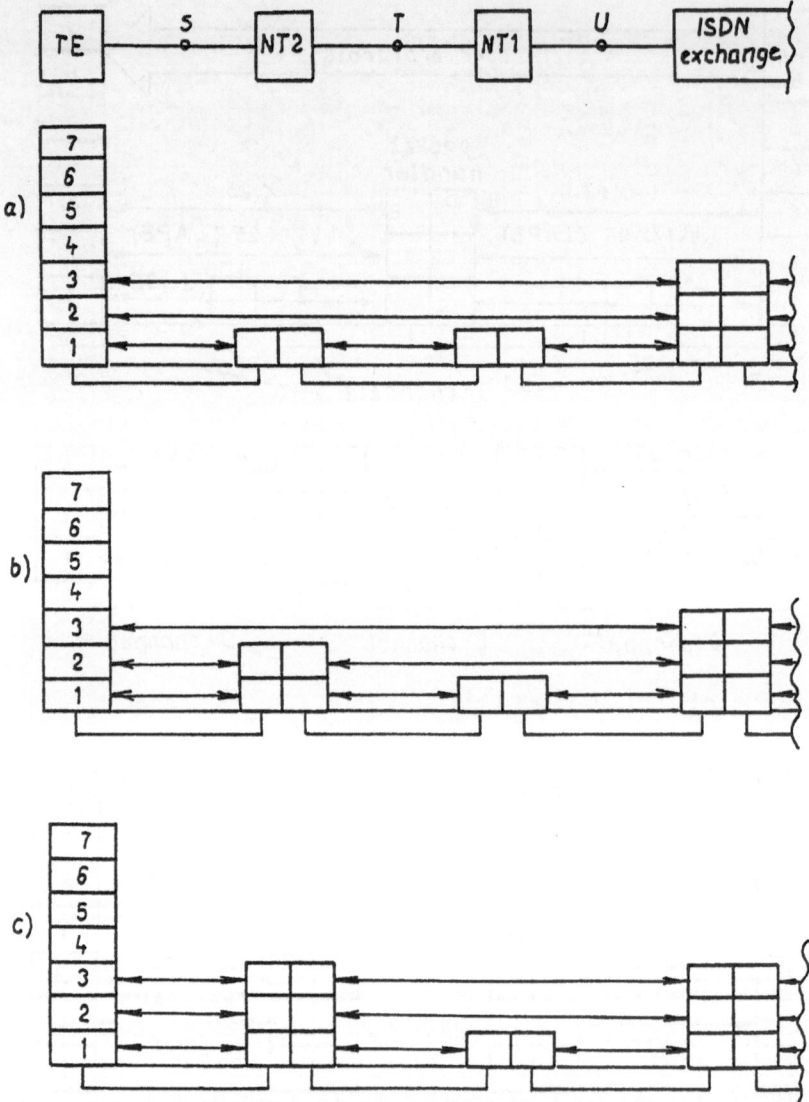

Fig. 4.16. Examples of access to an ISDN through the NT2 in the role of **a** an active bus; **b** a statistical multiplexer; **c** a private branch exchange.

From the point of view of the two main classes of services in the layered architecture the interworking may be one of two sorts:

- At the transmission capability level, that is, in the three lowest layers
- At the communication capability level, in the transport level or higher

This is shown in Fig. 4.17. In the former case (Fig. 4.17a) only data transmission services are provided, depending upon the equipment of the

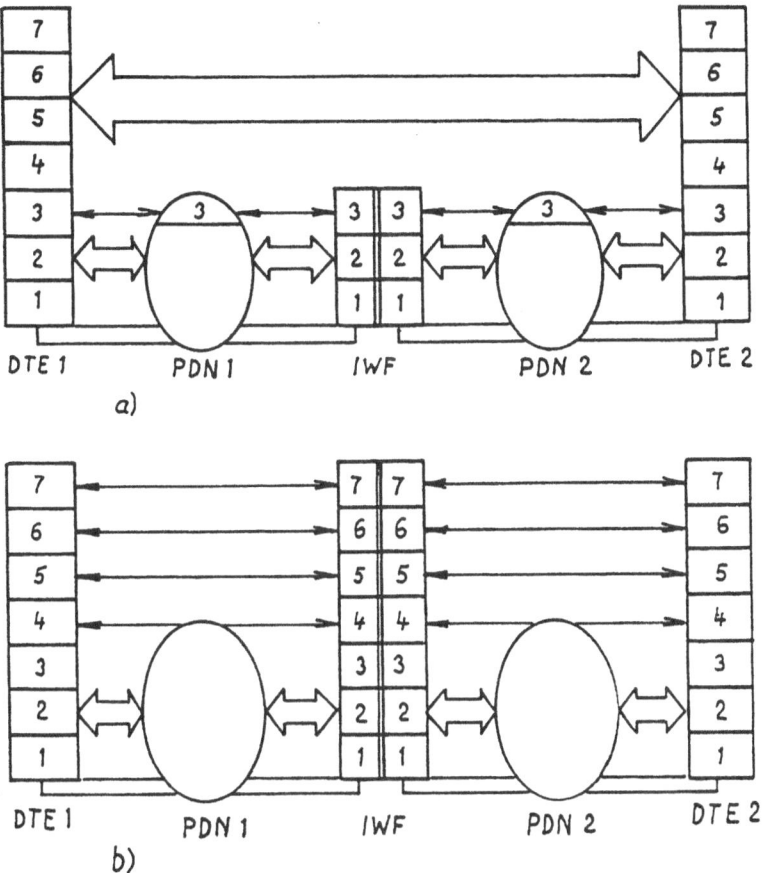

Fig. 4.17. Examples of interworking. **a** At the transmission capability level; **b** at the communication capability level.

network layer and the layers beneath it. For example, if a CSPDN and an ISDN providing circuit switched transmission services interwork, the resulting network tandem provides network services below the OSI network services. After adding an appropriate convergence protocol the services are enhanced up to network services explicitly defined in the reference model. The latter case (Fig. 4.17b) is applied if some modifications are needed in higher layers. The access to the PSPD via the PSTN and the PAD is an example of this.

Interworking at the transmission level recognizes two categories:

● Interworking by call-control mapping when all information involved in the call establishment command of one network is mapped (transformed) into information in the call establishment command of the network protocol of the other network (for example, interworking CSPDN–CSPDN and CSPDN– ISDN)

Table 4.2. Interworking between networks and some relevant CCITT recommendations

Network	CSPDN	PSPDN	ISDN	Mobile	Private
CSPDN		X.322	X.321		
	X.60, X.61, X.70, X.71, X.80	X.75, X.82	X.81	X.75	
PSPDN	X.322	X.323	X.325	X.324	X.327
	X.75, X.82	X.75	X.75	X.350, X.351 X.352	X.25
ISDN	X.321	X.325	X.320	X.324	
	X.81	X.75	X.75		

Upper documents refer to interworking arrangements; lower ones to interworking protocols.

- Interworking by port access when a required connection is established first in one network to the interworking unit and is extended over the other network, withholding all the necessary information in the call establishment command (for example, interworking PSTN– PSPDN)

The issue of interworking in general as well as in specific configurations is much more comprehensive and beyond the scope of this book. Problems associated with addressing are dealt with in Section 6.6 and those with quality of service in Section 7.1. Table 4.2 presents related CCITT documents by showing references for interworking arrangements and corresponding protocols valid for most network pairs.

4.5 Physical Layer [8,17,23,33,42,45,47,52,54,58,72]

4.5.1 General

This section slightly deviates from the others and also from the general design of this book. The following description of physical layer characteristics is based on the classical approach to interfaces and multiplexers, which is not fully in accordance with the OSI reference model. The tangible interface does not yet form the physical layer but it does provide for mechanical and electrical or optical interconnection. On the other hand, the physical layer protocol is essentially of a procedural nature. The inclusion of static multiplexing and fixed assignment of channels among physical layer features is not surprising because it follows previous chapters. Even though static multiplexing is not directly a physical layer function it provides data circuits on physical media and thus needs appropriate procedures. The CSPDNs and ISDNs based upon time switching employ these procedures as well, though this fact is usually not explicit in switching system descriptions.

   This section surveys the hitherto valid standards and, even if it cannot offer alternatives for them, it might serve as a guide and encourage further study.

   The physical layer of the OSI reference model is directly bound up with methods of transmission and is designated for activation, maintenance and deactivation of connections within these means. The transmission means is a data circuit based upon an arbitrary medium or media (wire/wireless, metallic/optical), and upon an arbitrary configuration and structure (private/ public, leased/switched). The term "physical" does not only imply a material substance. A circuit formed by bit strings in TDM is a physical medium too. Remember the wine-bowl shape of the standardization activity within OSI (Table 4.1), where standard documents relating to the physical layer form its large base. Each implementation of a new medium or communication configuration results in the elaboration of additional protocols or enhancement of the current ones, and this process is unlikely to be completed, particularly in the present age of rapid development of information transmission tools.

   Fig. 4.18 shows a simple example of a transmission path composed of two different cables: with copper pairs and with optical fibres. The physical layer accommodates the two media for data transmission and provides the upper layers with unified services (classes of service). Thus the physical layer hides the actual components in a black box invisible to users.

   The physical layer differs substantially from the other layers by the fact that below it there is a passive environment providing no services. The transmission medium (or a series of several media) is capable of transmitting strings of bits (serial transmission) or strings of groups of bits (parallel transmission) in one direction (simplex) or two directions (half- or full-duplex) between two points (point-to-point) or several points (multipoint, bus, ring). The physical layer activates one or several connections between service access points, provides for transparent data transmission – that is, one that is independent of data content and arrangement (sequence) – and deactivates the connection upon request from user or provider. The physical layer does not differentiate between the connection-mode and connec-

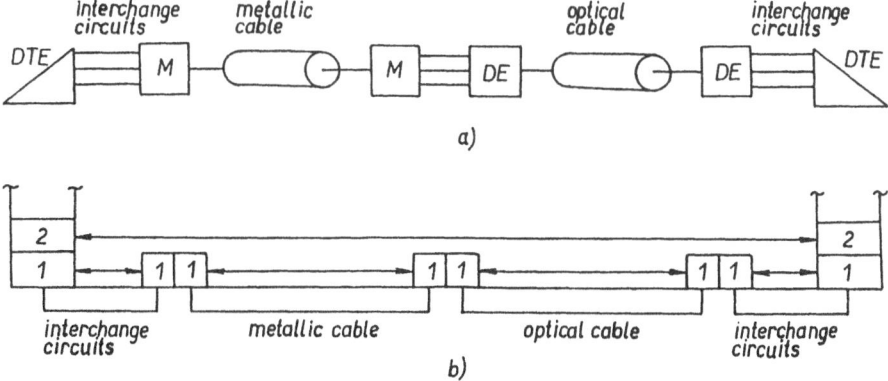

Fig. 4.18. An example of interconnection of two different means. **a** Physical diagram; **b** logical diagram.

tionless-mode of communication; this is the role of the upper layers of the reference model.

Performance (quality of physical services) depends upon the transmission media used, but its values need not be guaranteed during the duration of the physical connection.

The history of the physical layer specification is long and goes back to the 1960s. Of course, this protocol was known under the label "DTE–DCE interface" as a result of efforts to harmonize producers of data communication equipment (modems for example) and data processing equipment (computers, terminals). For practical reasons a boundary between DTEs and DCEs was set up which has delimited the competence of data service users and providers (mostly PTT administrations). Such a boundary, often designated as I2 (interface No.2 as distinct from the I1 interface between a DCE and a transmission circuit and I3, between a peripheral device and the control unit within the DTE) has enabled the unambiguous compatibility of equipment of various types.

The CCITT started the standardization activities in this area. The first CCITT recommendation was V.24, approved in 1964, followed by further V-series recommendations for data transmission over the PSTN and analogue leased lines, and later by the X-series for the PDNs. Some of them have appeared in the V-series as well as in the X-series (for example, V.10 is the same as X.26). Nevertheless, the definition of physical services prepared as a draft X.21C for the VIIIth CCITT Plenary Assembly in 1984 was finally approved four years later as X.211. This document definitely solved the location of synchronization and phasing control and assigned the former to the physical layer, while frame phasing and error control reside in the link layer. The synchronization refers to bits which form physical protocol data units (in the case of serial transmission). If, however, serial transmission of multiplexed bit streams is provided, as in time division systems or the ISDN, it refers to the contents of time slots, say octets, too.

4.5.2 *Interfacing*

Since the beginning of interface standardization, four characteristics have been taken into account:

- Mechanical
- Electrical
- Functional
- Procedural

This approach has survived in the OSI era, but the mechanical and electrical characteristics of physical layer protocols are specific only for the lowest layer which directly employs physical media and are beyond the scope of the OSI. Despite this, we consider all of these characteristics.

The basic ISO and CCITT documents dealing with the four characteristics are listed in Table 4.3. Six connectors are specified, but two of them are preferred for public digital services:

- The 15-pin connector for digital lines between DTEs and DCEs in PDNs
- The 8-pin connector for digital lines between terminal equipment and network terminations in the ISDN

The two connectors are schematically represented in Fig. 4.19. The other connectors are used in data transmission over analogue circuits:

- The 25-pin connector intended for serial and parallel modems, for telegraph signal converters in leased and switched circuits
- The 37-pin and 9-pin connectors largely used with serial modems, particularly if the backward channel is provided (separate 9-pin connector)
- The special (due to its different shape) 34-pin connector applied in the modems which equip broadband (48 kHz) telecommunication circuits

The main electrical characteristics significant for PDNs are shown in Table 4.4. Besides electrical values assigned to logical "0" and logical "1", the table gives an indication of the maximum signalling rate as limited by the length of

Table 4.3. Characteristics of physical layer protocols

Mechanical	Electrical	Functional	Procedural
15-pin connector ISO 4903	V.10/X.26, V.11/X.27	X.24	X.20, X.21
25-pin connector ISO 2110	V.28	V.24	X.20bis, X.21bis
34-pin connector ISO 2593	V.28, V.35	V.24, V.35	X.21bis
37-pin connector 9-pin connector ISO4902	V.10/X.26 V.11/X.27	V.24, V.36, V.37	X.20bis, X.21bis
8-pin connector ISO 8877	I.430	I.430	I.430

Table 4.4. Electrical characteristics of the DTE–DCE interfaces in PDNS

Type of circuit	Field of application	Electrical values	CCITT recommendation
Unbalanced double-current	Interconnection of integrated circuit equipments up to 100 kbit/s for 10 m (1 kbit/s for 1000 m)	0: $\geq$ +0.3 V 1: $\leq$ −0.3 V	V.10 X.26
Balanced double-current	Interconnection of integrated circuit equipments up to 10 Mbit/s 10 m (100 kbit/s for 1000 m)	0: point A has higher potential than B by $\geq$ 0.3 V 1: vice versa	V.11 X.27
Unbalanced double-current	Interconnection of equipments based on discrete elements up to 20 kbit/s for 15 m	0: +3 V 1: −3 V	V.28
Balanced double-current	Interconnection of high rate equipments up to 48 kbit/s	0: point A has higher potential than B by 0.55 ± 0.11 V 1: vice versa	V.35

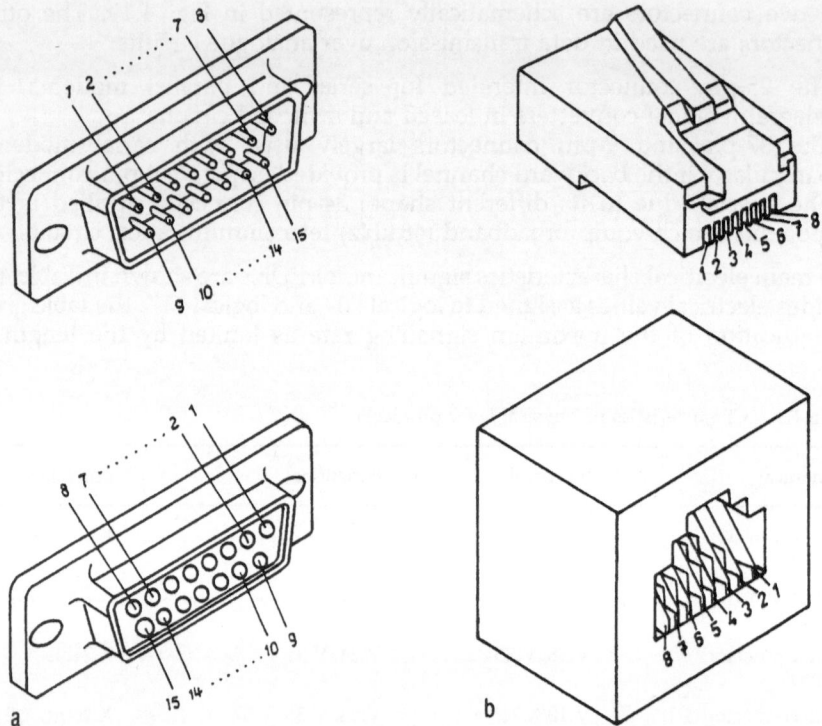

Fig. 4.19. Examples of two standard connectors. **a** Plug and socket of the 15-pin connector for PDNs; **b** plug and socket of the 8-pin connector for the ISDN.

interchange circuits used (note that this rate decreases approximately linearly with increasing length). The V.10/X.26 and V.11/X.27 characteristics designed for general use with integrated circuit equipment are the most important for subscriber lines of PDNs. The corresponding recommendations are often called IC (integrated circuit) recommendations. Balanced interchange circuits are characteristic of higher protection against noise and hence they operate with higher signalling rates over long distances. The penalty for this advantage is the requirement for two wires (and two connector pins) per interchange circuit.

Functional characteristics refer to the number of interchange circuits and their functions. The richest set is given in CCITT V.24. This recommendation was specified for the modem line, also standardized in the CCITT V-series recommendations. The number of 25 interchange circuits set for modems in 1964 has increased to 40 (of the 100-series plus 12 for automatic calling of the 200-series) in accordance with DCE requirements and with PTT maintenance policy. The selected interchange circuits applied to analogue access to PDNs by means of modems are represented in Table 4.5 together with the assignment of pins in standard connectors. The circuit designation is taken from CCITT V.24, the numbering (lettering) is based on corresponding ISO standards.

Table 4.5. Selected interchange circuits between the DTE and DCE in PDNs (CCITT V.24)

Function	Name of circuit	Direction (from–to)	Designation	Connector pin allocation 25-pin	37-pin	34-pin
Ground	Signal ground or common return		102	7	19	B
Data	Transmitted data	DTE–DCE	103	2	4,22	P,S
	Received data	DCE–DTE	104	3	6,24	R,T
Control	Request to send	DTE–DCE	105	4	7,25	C
	Ready for sending	DCE–DTE	106	5	9,27	D
	DCE ready	DCE–DTE	107	6	11,29	E
	Connect DCE to line	DCE–DTE	108/1	20	12,30	
	DTE ready	DTE–DCE	108/2	20	12,30	
	Data channel received line signal detector	DCE–DTE	109	8	13,31	F
	Calling indicator	DCE–DTE	125	22	15	
Timing	Transmitter signal element timing	DCE–DTE	114	15	5,23	Y,AA
	Receiver signal element timing	DCE–DTE	115	17	8,26	V,X
Testing	Loopback/maintenance test	DTE–DCE	140	21	14	
	Local loopback	DTE–DCE	141	18	10	
	Test indicator	DCE–DTE	142	25	18	

Since the V.24 interface for PDNs is unnecessarily complex and there is no need to convey each command/response over a separate circuit, a simpler set of interchange circuits has been proposed for the PDN environment (Table 4.6). The control part has been reduced to two circuits, one for each direction, while for the timing four circuits have been adopted. Note that the 15-pin connector is preferred to the 9-pin connector because of the balanced double current which requires two wires for each interchange circuit (see the pairs of pins assigned to each circuit in Table 4.6). If the unbalanced double current is used, only one pin per interchange circuit is chosen together with the common ground (pin 8).

The most modern transmission and switching means – the ISDN – further simplifies this: only two information transfer circuits are sufficient for the exchange of bit strings between an item of terminal equipment (TE) and its network termination (NT) carrying all three channels (2B + D). The circuits for power supply are optional: they are designated for distant current supply of TE by NT or vice versa (Table 4.7). For further details, see CCITT I.430.

CCITT recommendation V.24 also contains interchange circuits for automatic calling and answering units if included in the modem. Table 4.8 presents a summary of the so-called 200-series interchange circuits where four

Table 4.6.   Interchange circuits between the DTE and DCE in PDNs (CCITT X.24)

Function	Name of circuit	Direction (from–to)	Designation	15-pin connector
Ground	Signal ground or common return		G	8
	DTE common return		G_a	
	DCE common return		G_b	
Data	Transmit	DTE–DCE	T	2,9
	Receive	DCE–DTE	R	4,11
Control	DCE control	DTE–DCE	C	3,10
	DCE indication	DCE–DTE	I	5,12
Timing	Signal element timing	DCE–DTE	S	6,13
	Byte timing	DCE–DTE	B	7,14
	Frame start identification	DCE–DTE	F	7,14
	DTE signal element timing	DTE–DCE	X	7,14

Table 4.7.   Interchange circuits between terminal equipment (TE) and network termination (NT) in ISDNs

Function	Direction (from–to)	Connector pin allocation
Information transfer	TE–NT	3,6
Information transfer	NT–TE	4,5
Power supply	TE–NT	1,2
Power supply	NT–TE	7,8

Table 4.8. Interchange circuits between the DTE and DCE for automatic calling (CCITT V.24)

Name of circuit	Direction (from–to)	Designation	Connector pin allocation
Signal ground or common return		201	7
Call request	DTE–DCE	202	4
Data line occupied	DCE–DTE	203	22
Distant station connected	DCE–DTE	204	13
Abandon call	DCE–DTE	205	3
Digit signal 2^0	DTE–DCE	206	14
Digit signal 2^1	DTE–DCE	207	15
Digit signal 2^2	DTE–DCE	208	16
Digit signal 2^3	DTE–DCE	209	17
Present next digit	DCE–DTE	210	5
Digit present	DTE–DCE	211	2
Power indication	DCE–DTE	213	6

digital signals correspond to the bit positions of the binary coded decimal digit of the dialled number. These interchange circuits support the call control protocol (CCITT V.25) for data transmission over the PSTN and for access to PDNs through this network. Beside the V.25 protocol, V.25bis also allows the use of interchange circuits of the 100-series for this purpose.

The procedural characteristics are closely related to the network protocol used (CCITT X.20 for start–stop DTEs and X.21 for synchronous DTEs) and, as these protocols were elaborated before the OSI reference model was established, their separation is difficult. Comparing the protocols for leased and switched data circuits (for the former the network layer is transparent) the connection activation is opened by the transition of DTE control circuit (C) and the DCE indication circuit (I) to the ON-condition (in the ready state both circuits are in the OFF-condition), which is followed, in synchronous transmission, by a sequence of synchronous characters. We shall return to this problem in Section 4.7.2.

4.5.3 Multiplexing [22,34,40,44]

Multiplexing is the process of assembling a number of channels into a common bearer channel of higher transmission capacity. The inverse process, involving disassembly by division of the common channel into the original channels is *demultiplexing*. The system of multiplexing with corresponding demultiplexing is called a multiplex system. As explained briefly in Section 3.1, multiplexing is a method of economic utilization of the transmission media, by overhead lines, cables with metallic conductors or optical fibres or by radio links terminated by radio transmitters and receivers, including those which are located on telecommunication satellites or in mobile stations. The basic channels provided by these media usually yield a much greater transmission capacity than is needed for data transmission between communicating DTEs. The transmission capacity is given either by the available frequency band or the maximum data signalling rate (bit rate) of this channel. For a channel of defined quality there is an unambiguous correspondence between the former and the latter.

The uncoordinated development of multiplex transmission technology has brought about a certain inconsistency in multiplex terminology. The dominant term "multiplex" implies the assembly of narrow-band or slow-speed channels into a broadband or high-speed channel, respectively. On the other hand, the designation of both kinds of multiplexes, namely "frequency division multiplex (FDM)" and "time division multiplex (TDM)", is related to channel disassembly. FDM has been known in telegraph transmission as voice frequency telegraphy (VFT) because of its utilization of the voice grade channel (telephone channel) for multichannel telegraph transmission. In telephone transmission we speak rather of multichannel carrier systems because of the role of the carrier wave which enables the transposition of the telephone signal into a higher frequency position. Both of these original multiplex systems are characteristic of analogue signal processing.

Nowadays when speaking of multiplex systems we usually mean TDM systems employing digital transmission. In any multiplex system, a multiplexer is used for channel assembly onto the multiplex channel and its reverse – a demultiplexer – for channel disassembly to the original channels. For both-way transmission a multiplexer must be coupled with a demultiplexer which processes signals proceeding in the opposite direction. In this case we can speak of the multiplex circuit and tributary circuits and use the term *muldex* (a term of a similar etymology to "modem") for the multiplexer and demultiplexer coupled together to assemble tributary circuits onto a bearer circuit and vice versa. Fig. 4.20 illustrates these fundamental terms. A multiplex is homogeneous if all tributary channels have the same parameters, and non-homogeneous if the tributary channels mutually differ, by such things as data signalling rate (or bit rate), transparency and (in the case of untransparency) character format. The overall bit rate of the bearer channel of a multiplex is approximately the sum of the bit rates of the tributary channels. The term "approximately" means that the bit rate of the bearer channel is always a little higher than the product or sum because some extra information

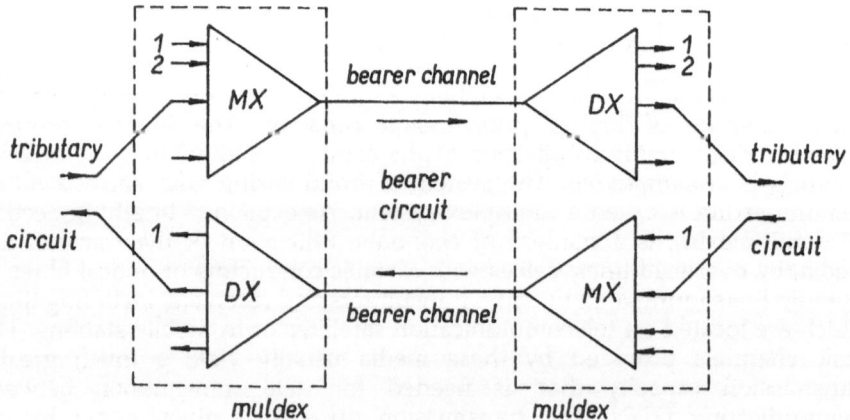

Fig. 4.20. Basic multiplexing principles and terms. DX, demultiplexer; MX, multiplexer.

must be transferred for the establishment and maintenance of the alignment of the phase difference between multiplexer and demultiplexer. This difference can be explained by the necessary angular difference between a rotating switch which produces time slots by scanning the input channels (thus representing a multiplexer) and another, distant, switch rotating with the same angular velocity and distributing the signals to the output channels (representing a demultiplexer) when the multiplex channel introduces a certain transmission delay.

The establishment of the correct angle can be made by selecting a channel specialized for the transmission of a signal which sets or readjusts the arm of the demultiplexer switch into a certain agreed position. During a complete cycle of the multiplexer or demultiplexer every channel is allowed to achieve an elementary transmission in turn. The signals transmitted on the bearer channel during this cycle constitute a *frame*.

The movement of the multiplexer's and demultiplexer's arms can be continuous (in a synchronous multiplex system) or can stop after performing the cycle (asynchronous system, perfected in start–stop transmission systems). Multiplex systems in PDNs utilize synchronous multiplexing. The pace of the multiplexer's and demultiplexer's arms is controlled by the multiplexer and demultiplexer clocks, respectively. The clock is represented by a periodic timing signal derived usually from a stable frequency generator. Any divergence of clock frequencies between a multiplexer and the corresponding demultiplexer would inevitably lead to intolerable misalignment with the risk of sending signals to the wrong demultiplexer outputs. Therefore clock information is derived from the received multiplex signal or provision is made in the network for a suitable distribution of clock signals.

As to the portion of data transmitted by a channel in a multiplex system during one cycle we speak of bit interleaving, character interleaving and byte interleaving. The preference for bit interleaving can be explained by the fact that this interleaving guarantees the shortest signal delays.

A summary of CCITT recommended TDM systems utilizable in CSPDNs is given in Table 4.9. Muldexes for the transmission of telegraph signals and anisochronous data signals whose use should be considered for start–stop type DTEs (classes 1 and 2 according to X.1, see Table 2.1) are also included.

For synchronous classes 3 to 6 with data signalling rates of 600, 2400, 4800, and 9600 bit/s, multiplex systems according to recommendations X.50 and X.51 are used. The data is inserted into the multiplex aggregate via the multiplexer and distributed into the channels via the demultiplexer. Channel interleaving can be achieved by high-density sampling (in the case of telegraph multiplexers according to R.111), by bits (in telegraph-type multiplexers according to R.101) or groups of bits (in data multiplexers according to X.50 and X.51). A relevant group of overhead bits within a frame serves to maintain frame alignment.

The group of data (information) bits I transmitted by the data multiplexer during a time slot according to X.50 or X.51 is complemented by two bits serving in-channel purposes. It is status bit S that is used for control purposes and bit A for indication and alignment. These two auxiliary bits with provision for a fixed number of data bits (information bits) can be compared with an envelope for a letter – hence CCITT recommendations X.50 and X.51 use the term *envelope*.

Table 4.9. CCITT recommended TDMs utilizable in PDNs

CCITT recommendation	Denomination of multiplex system	Transparence	Regeneration	Bit rate (kbit/s) Bearer channels	Tributary channels	Number of tributary channels
X.50	Multiplex for interface between synchronous data networks	Yes[a]	Yes	64	0.6 2.4 4.8 9.6	80 20 10 5
X.51	Multiplex for interface between synchronous data networks, time envelope 8 + 2	Yes[a]	Yes	64	0.6 2.4 4.8 9.6	80 20 10 5
R.101	Code and speed dependent TDM system for anisochronous telegraph and data transmission with bit interleaving	No	Yes	2.4	0.05 0.075 0.1 0.11 0.2 0.3	46 30 22 22 15 7
R.111	Code and speed independent TDM system for anisochronous telegraph and data transmission	Yes	No	64	0.05 0.1 0.2 0.3 0.6 1.2	240 120 60 60 30 15

[a] Bit sequence independence.

The X.50 multiplex uses an 8-bit envelope which has the structure AIIIIIIS, and in CCITT terminology it is called 6 + 2 envelope. The X.51 envelope occupies a 10-bit slot with the structure SAIIIIIIII and in CCITT terminology it is designated 8 + 2 and known as 10-bit envelope. If a certain data flow (without S and A bits) should have data signalling rates standardized for user classes of service in PDNs according to X.1, that is, 600, 2400, 4800, and 9600 bit/s, the data channel in the multiplex channel of the 6 + 2 envelope system must have a data signalling rate elevated by 33 per cent (to 800, 3200, 6400 and 12 800 bit/s) and that in the 8 + 2 envelope system by 25 per cent (to 750, 3000, 6000 and 12 000 bit/s).

The advantage of the 6 + 2 envelope system is that for circuit switching it is possible to use the same switching system as for the digital telephone exchange on PCM basis. The advantage of the 8 + 2 envelope system is that it corresponds to the current octet structure of data transmitted and received by DTEs and the A bit can be used directly for byte timing. In principle, the possibility of interworking between both systems is required. The conversion facility extracts data bits out of the envelopes of one type and inserts them into the envelopes of the other type. This requirement is imperative for the choice of frame size and framing structure. The number of information bits (data bits) transmitted by either system within a unit of time is the same, as is the capacity expressed by the number of channels of the homogenous configuration.

In a system according to X.50 the frame contains 80 slots enveloped as 6 + 2 (640 bits). Therefore, in a 64 kbit/s bit stream 100 frames per second are transmitted. The elevated rate channel 12.8 kbit/s (the elevation being caused by introducing A and S bits into the flow of D bits) employs every fifth envelope slot and has 16 envelope slots in each frame (see the part of the frame with four time slots designated A, B, C, D in Fig. 4.21). The capacity of the homogeneous system is then five channels of 9.6 kbit/s, ten channels of 4.8 kbit/s, twenty channels of 2.4 kbit/s, or eighty channels of 600 bit/s. The system also allows for the provision of non-homogeneous configurations with simultaneous existence of channels with various bit rates under the condition that their sum does not exceed 48 kbit/s.

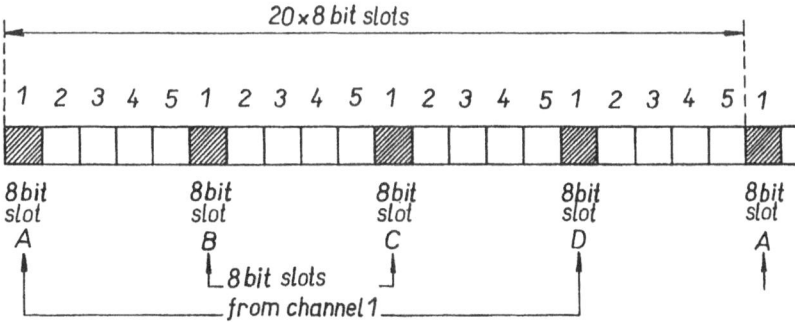

Fig. 4.21. Formation of the 9.6 kbit/s data channels from 6 + 2 envelope slots in the 64 kbit/s bearer channel according to CCITT recommendation X.50.

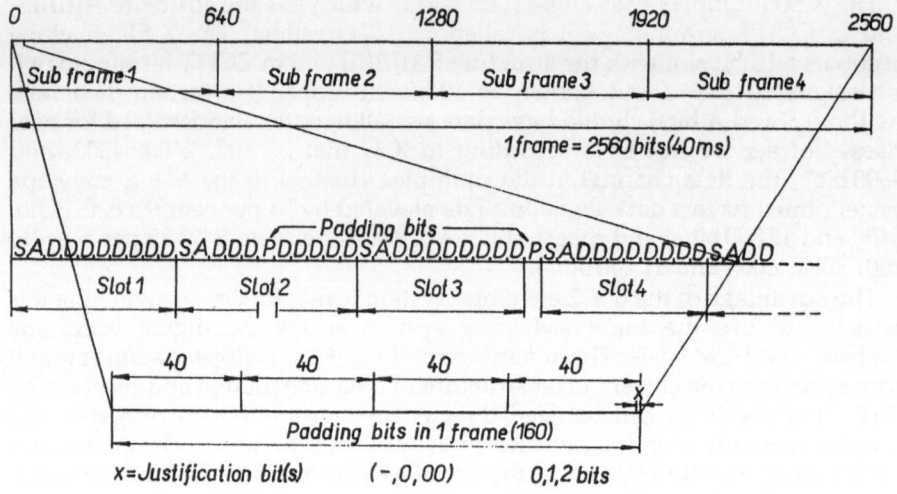

Fig. 4.22. Frame structure of the X.51 multiplex with 8 + 2 envelope slots. D, data bit.

In the system according to X.51 the frame consists of 2560 bits and occupies a time interval of 40 ms. Its structure is illustrated in Fig. 4.22. There are 160 padding bits P altogether, equally distributed over the frame, therefore every 16th bit is a P bit. The P bits are utilized for frame alignment and in various other functions, such as error control and interworking between the multiplexer and distant demultiplexer. Within a frame, the 14-bit frame alignment word is repeated four times. This subdivides the frame into four subframes. The last of the 160 padding bits is a justification bit. By preserving or leaving out the justification bit we have the possibility of coping with the data signalling rate variation within a span of 4×10^{-4}.

The net bearer data signalling rate for the transmission of envelope slots 8 + 2 is 60 kbit/s. This stream will accommodate five elevated speed channels of 12 000 bit/s, that is, five data channels of 9600 bit/s, which is the same capacity as the X.50 multiplex. A similar consideration applies to other data signalling rates.

For anisochronous (see Appendix 1) operation in user classes 1 (300 bit/s) and 2 (up to 200 bit/s) the use of the following transmission methods and means is internationally standardized:

- Inserting isochronous (see Appendix 1) signals into an isochronous user bit stream of 600 bit/s or – when this is not available – 2400 bit/s according to CCITT recommendation X.52. The principle of this insertion is evident from Fig. 4.23. Start–stop receiver (SSR) provides samples (b) from the centres of unit intervals of the received start–stop character signal (a) and passes them into the isochronous signal transmitter IT. The distant isochronous signal receiver IR passes these samples to its start–stop transmitter which produces the desired regenerated start–stop character signal with a certain time delay but with the original modulation rate

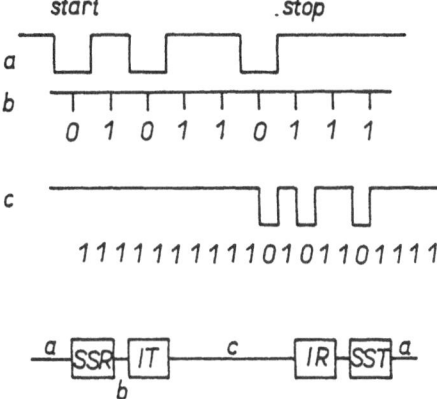

Fig. 4.23. The principle of the transmission of start–stop signals into an isochronous bit stream of higher data signalling rate. **a** Input and corresponding output start–stop signal; **b** sequence of samples at start– stop receiver output; **c** isochronous bit stream. IR, isochronous receiver; IT, isochronous transmitter; SSR, start– stop receiver; SST, start–stop transmitter.

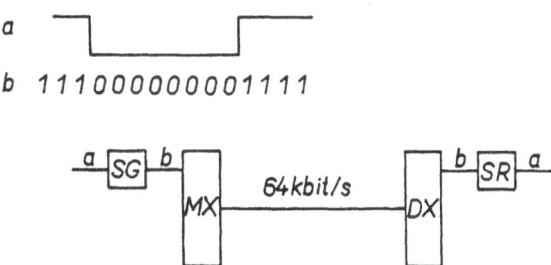

Fig. 4.24. The principle of transparent time division multiplex with multiple sampling according to CCITT recommendation X.111. **a** Input and the corresponding output binary signal; **b** sequence of samples inserted into the 64 kbit/s bearer bit stream on the multiplexer side and extracted at the demultiplexer side. DX, demultiplexer; MX multiplexer; SG, sampling generator; SR, signal restitutor.

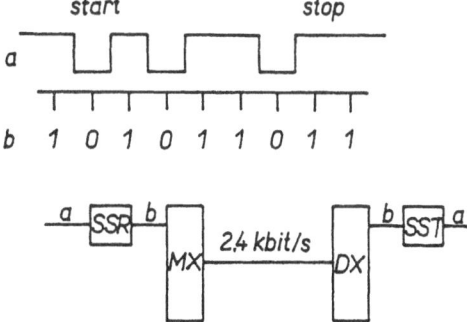

Fig. 4.25. The principle of the R.101 time division multiplex system for regenerated transmission of start–stop signals. **a** Input and regenerated transmission of start–stop signal; **b** stream of samples inserted into the 2.4 kbit/s bearer.

- Code and speed transparent channels of 200 baud in voice frequency telegraph systems according to R.37 and R.38 for user class 2 from X.1.
- Code and speed transparent channels of the TDM system according to R.111 for anisochronous telegraph and data transmission by periodical sampling with sampling rates substantially higher than the modulation rate of the transmitted signal (Fig. 4.24). The stream of samples (b) from the input signal (a) is inserted into the 64 kbit/s bearer and extracted at the demultiplexer to acquire its original form (a). The capacity of the multiplex system is sixty data channels of 200 bit/s channels with 5 per cent of inherent distortion or the same number of data channels of 300 bit/s with 7.5 per cent inherent distortion
- Code and speed non-transparent start–stop channels of the multiplex system with signal regeneration for anisochronous telegraph and data transmission using bit interleaving according to R.101. The principle consists of the transmission of samples (b) of the received start–stop signal (a) via the bearer stream 2.4 kbit/s into distant start–stop transmitters (with signal (a) delayed and regenerated) at the output of the demultiplexer (Fig. 4.25). The capacity of this system is ten channels of 200 bit/s or seven channels of 300 bit/s

Another field of application of TDM for data transmission services is the insertion of signals from DTEs into the B- and D-channels (and vice versa – extraction for DTEs from B- and D-channels) of the basic and primary access to the ISDN.

The basic access (defined by CCITT recommendation I.430) yields in each direction (TE–NT and NT–TE) two B-channels (basic channels B1 and B2) of 64 kbit/s and one D-channel of 16 kbit/s. The frame (Fig. 4.26) of the basic access consists of 48 bits and covers a time interval of 250 μs. This means that any single bit position within this frame has a repetition rate of 4 kbit/s. The B-channel bits are grouped into octets. In the frame there are two B1 octets, two B2 octets and four uniformly distributed D-channel bits. The D-channel bits transmitted from the DTE are echoed back by the NT to the DTE as E bits.

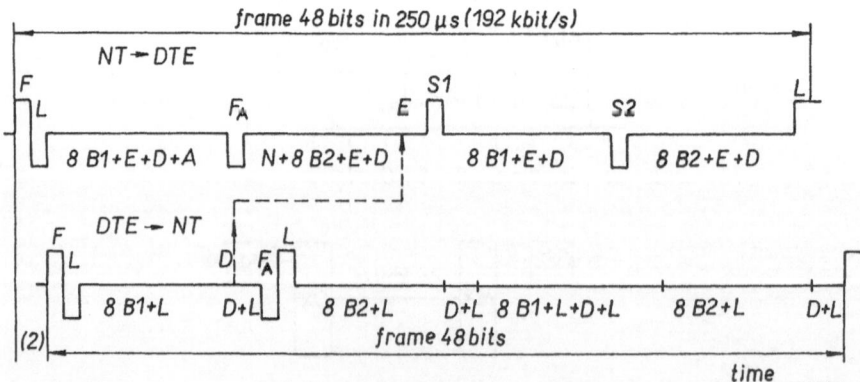

Fig. 4.26. Frame structure of the time division multiplex system for multiplexing and signalling rate adaptation in the ISDN basic access circuit.

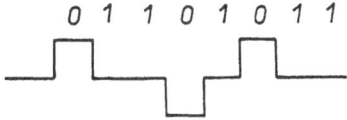

Fig. 4.27. Pseudobinary signal with elimination of direct current component.

Every frame contains one framing bit F and one auxiliary framing bit FA and its negation N. Bit A is used for the activation protocol, bit M is reserved for multiframing purposes. The direct current balancing bits L serve to eliminate the undesired direct current component which could appear due to the use of the pseudobinary alternate mark inversion code where "0" is represented by either a positive and a negative pulse (they alternate) and "1" by the absence of line signal (see Fig. 4.27). The use of bit S is not yet standardized.

The primary rate user–network interface (CCITT I.431) provides time division of the primary bit rate, 1544 kbit/s or 2048 kbit/s. This time division uses the principles laid down in the CCITT G.700 series of recommendations (general aspects of digital transmission systems) and is primarily intended for digital transmission of speech signals of the telephone channel utilizing the band 300–3400 Hz. The application of the sampling theorem to the telephone signal brought about the use of the 8 kHz sampling frequency which was decisive for the choice of the 125 μs frame. The requirement for eight bits necessary for carrying information about sample values then resulted in the requirement for 64 kbit/s for every telephone channel. The 1544 kbit/s system then yields 24 channels of 64 kbit/s and the 2048 kbit/s system 32 channels of 64 kbit/s. The frames consist of 193 and 256 bits, respectively.

The ultimate trend for data transmission over the ISDN involves the use of such DTEs which could be directly connected to the user–network interface in reference points S/T so as to be consistent with the above mentioned rules of basic and primary access. However, the use of PDN-type DTEs or of DTEs designated for interfacing to telephone-type modems necessitates the insertion of terminal adaptors (TAs). Multiplexing 8, 16 and 32 kbit/s bit streams on one aggregate 64 kbit/s channel involves assigning combinations of bits (each of them representing 8 kbit/s) to the lower speed channels. If only one of the lower bit rates is needed, we speak about rate adaptation. The principles of multiplexing and rate adaptation are laid down in CCITT recommendation I.460. Suppose a B-channel octet (repeated in every frame) in the basic access or primary access has its bit positions designated 12345678. An 8 kbit/s stream is allowed to occupy any bit position, a 16 kbit/s channel any of the positions 12, 34, 56 or 78, a 32 kbit/s stream occupies positions 1234 or 5678. All unused bits are set to binary "1". The adaptation for DTEs is made in terminal adaptors.

According to CCITT X.30 a terminal adaptor adapts PDN user class bit rates in two stages (Fig. 4.28): in addition to the adaptation of 64 kbit/s down to 8 kbit/s or 16 kbit/s (generally $2^k \times 8$ kbit/s) described above there is a lower stage for the adaptation of $2^k \times 8$ kbit/s to the bit rates of synchronous user classes 600, 2400, 4800 and 9600 bit/s. The adapting function is a by-product of the multiplexing function based on the use of frames and multiframes.

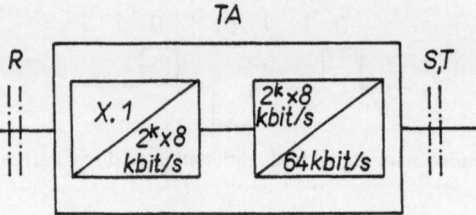

Fig. 4.28. Two-stage signalling rate adaptation in an ISDN terminal adaptor (TA).

Table 4.10 illustrates the specific solution of establishing two 2400 bit/s channels on a 8 kbit/s bearer. To this aim the 8 kbit/s stream is divided into octets, five consecutive octets form a frame, two frames (containing 80 bits in total) form a multiframe. Each bit position has a bit rate of 100 bit/s. To achieve a bit rate of 2400 bit/s it is sufficient to select 24 bit positions in the multiframe and assign them a channel. In the example of Table 4.10, bit positions assigned to data are designated P, Q and R and are framed into rectangles; one channel is marked by bit positions in bold (even numbered columns). Bits E carry supplementary signalling information about user signalling rate. Bits SP, SQ and SR are status bits, X is reserved for auxiliary functions. The TA with its adapting and multiplexing functions also serves as a signalling converter between the two interfaces located in reference points R and S/T.

Table 4.10. Multiframe for the insertion of two 2.4 kbit/s streams into an 8 kbit/s stream

Frame	Number of octet in frame	Number of bit in octet							
		1	2	3	4	5	6	7	8
Odd	0	0	0	0	0	0	0	0	0
	1	1	P1	P1	P2	P2	P3	P3	SP
	2	1	P4	P4	P5	P5	P6	P6	X
	3	1	P7	P7	P8	P8	Q1	Q1	SQ
	4	1	Q2	Q2	Q3	Q3	Q4	Q4	SQ
Even	0	1	1	1	0	E4	E5	E6	E7
	1	1	Q5	Q5	Q6	Q6	Q7	Q7	SR
	2	1	Q8	Q8	R1	R1	R2	R2	X
	3	1	R3	R3	R4	R4	R5	R5	SR
	4	1	R6	R6	R7	R7	R8	R8	SP

4.6 Link Layer [7,33,47]

4.6.1 Development of Link Protocols

The link layer is defined between adjacent systems (stations), for example: end system (user, subscriber) and nearest intermediate system (switching node, concentrator); switching node (exchange) and adjacent switching node; end system and end system in the case of the transparent link layer, etc. The link protocol serves the control of communication over a data link (hence its name) and has been indispensable from the very beginning of data transmission over point-to-point as well as multipoint data circuits, even though the control was not then in compliance with the layered architecture. Beside the physical protocols, it is the oldest protocol and its history deserves at least a few lines.

Functions executed by the link protocol (see Section 4.3) refer to the entire behaviour of a data link regardless of the transmission medium upon which it is built up. However, the kind and origin of means (cable – wired, radio and satellite – wireless, fibre optic), and their performance (quality of service of the physical layer) influence the choice of a link protocol or, at least, values of its parameters in order to meet the requirements of a network layer or a user. Thus, link protocols were designed not only by data communication and computer communication producers but also by PTTs and users. Some of these protocols, sufficiently sophisticated, were readily accepted by the standardization bodies, ISO and CCITT.

In the beginning the standardization activity dealt with alphabets as the basis for the operation of data processing equipments, peripherals and data transmission equipment. The International Telegraph Alphabet ITA2 (based upon the 5-bit Baudot code) was the first code adopted worldwide (see CCITT S.1). The appropriate 7-bit code is derived from the American standard US ASCII and was approved in 1967 by the ISO council, first as a recommendation, later as international standard ISO 646 (hence its name IS0–7). It appeared in the CCITT White Book in 1968 as recommendation V.3 under the name IA5 (International Alphabet No.5) or CCITT 5 and in 1984 it was transferred to the T-series recommendations as T.50. The IA5 comprises 128 8-bit combinations, ten of which are reserved for telecommunication control (see Table 4.11). These control characters form the basis for character-oriented formats and protocols.

In 1965 IBM worked out a link protocol (BSC – binary synchronous communication) for its terminal product line not only with the IA5 (more precisely US ASCII) but also with the 8-bit code EBCDIC. In 1968 the BSC was approved as American standard DLC (X3.16) and proposed to ISO for international use. ISO issued its basic mode in 1971 as recommendation 1745. Four years later it changed this recommendation into an international standard accompanied with complements forming further standards (2111, 2628, 2629) employing experience of ANSI X3.26 and ECMA (ECMA-16, 24, 26, 27, 28, 29, 37). We shall call this standardized link protocol ISO/BSC protocol in order to avoid confusion with other ISO protocols, even though there are slight differences between IBM and international versions.

In the meantime CCITT worked out within its former study group Special A a bit-oriented link protocol which was approved at the IVth Plenary Assembly

Table 4.11. Communication control characters of the IA5 (ISO-7, CCITT-5)

Designation	Significance	Code representation
SOH	Start of heading	1 0 0 0 0 0 0
STX	Start of text	0 1 0 0 0 0 0
ETX	End of text	1 1 0 0 0 0 0
EOT	End of transmission	0 0 1 0 0 0 0
ENQ	Enquiry	1 0 1 0 0 0 0
ACK	Acknowledgement	0 1 1 0 0 0 0
DLE	Data link escape	0 0 0 0 1 0 0
NAK	Negative acknowledgement	1 0 1 0 1 0 0
SYN	Synchronous idle	0 1 1 0 1 0 0
ETB	End of transmission block	1 1 1 0 1 0 0

in 1968 as recommendation V.41. The protocol, however, suffered from many limitations and it was again IBM who pioneered work in this area, developing the new bit-oriented link protocol SDLC in 1969. This protocol (more precisely, class of protocols) served as the background for work within the X3S3.4 Group of ANSI and resulted in the advanced data communication control procedure (ADCCP), a version of SDLC standardized in the USA. At the same time ECMA issued its versions numbered 40, 49, 60, 61 and 71, and together with ANSI proposed it as a draft of the high level data link control (HDLC) to ISO. The ISO working group TC 97/SC 6 worked on this protocol, which in turn was approved as a series of standards (3309 in 1976, 4335 in 1979, 7809 in 1984, etc.). It is not surprising that large computer communication manufacturers have also designed and implemented their own versions of HDLC. Nevertheless, after some revisions during the study periods, protocols LAP and LAPB (see later) which were eventually chosen and approved in 1980 and 1984 are almost identical with certain classes of HDLC specified in ISO 7809. Note, however, that even slight divergencies may cause incompatibility between equipments in operation.

This overview gives an illustration of the efforts that have been exerted for a relatively long time by strong groups of specialists in an apparently narrow area – one layer out of seven of the OSI reference model.

In the following section we describe briefly the principles of two link protocols, namely ISO/BSC and some versions of HDLC which play an indispensable role in PDNs and the ISDN.

4.6.2 Character-Oriented ISO/BSC Protocols [33,45,47,58]

This protocol seems to be obsolescent, mainly because of the following disadvantages compared with the bit-oriented HDLC:

- Code and data structure dependence
- Only half-duplex transmission and two-way alternate communication is controlled

- Inefficient error control
- Low exploitation of the transmission capacity of media
- Only centralized control is permitted

The ISO/BSC uses the character-oriented format based upon IA5. It guarantees the unique recognition of commands, responses and identifiers from user data unless they are in different codes. This is evident from the structure of frame format allowed in the ISO/BSC:

SYN SOH header STX data ETX (ETB) BCC

where the header may comprise control and identification data such as addresses, serial numbers and commands for frame processing. The field "data" contains arbitrary non-control IA5 characters; ETX terminates a data message (last data fragment) while ETB indicates the end of the data fragment unless it is the last one, and BCC – block-check character or characters – is generated by an error-correcting code (iterative, polynomial). The original BSC protocol has been elaborated for synchronous transmission (hence its abbreviation), but it is also adequate for start–stop transmission if each data character is preceded by one start bit and succeeded by one or two stop bits. Iterative coding adds another bit, the parity bit, to each character: odd parity for synchronous and even parity for start–stop transmission.

The recognition of data from the remaining part of the frame fails if the data contains any control character from Table 4.11, to which the receiving station would react as to a command. To prevent this, the transparent mode has been elaborated for code independent transmission. Code transparency is achieved by inserting DLE in front of all control characters (including DLE) within the frame. Thus only pairs of characters beginning with DLE are regarded as commands or responses with the exception of double DLE (DLE DLE) which retains the original meaning of data (one DLE must be cancelled after receiving a fragment).

The error control is performed by the ARQ stop-and-wait method with the A-scheme (ACK, NAK, time-out) which also enables control of data flows.

Since the control is strictly centralized, only one station, called the control station, is responsible for controlling communication with one or several tributary stations and for solving exceptional states which may occur in the process of communication. The communication is, however, two-way alternate, therefore both stations (control station as well as tributary station) have a right to send data (it becomes the master) and to receive data (it becomes the slave). The labelling of master/slave has to change during a communication according to the direction of user data flow. Two modes are therefore necessary: *selecting*, when the control station is at the same time master, and *polling* when it is a slave. Both modes are shown in Fig. 4.29.

In the selecting mode (Fig. 4.29a) the control–master station A sends a selecting command ("selecting", because one station among several tributary stations on the multipoint data circuit is selected by its address). After a positive response from the selected station B (say ACK) station A transmits data.

The polling mode is again initiated by the control station A, which is now the slave station waiting for data from a tributary–master station. After receiving the polling command station B transmits data (if any available), otherwise it sends a negative response, say NAK (Fig. 4.29b).

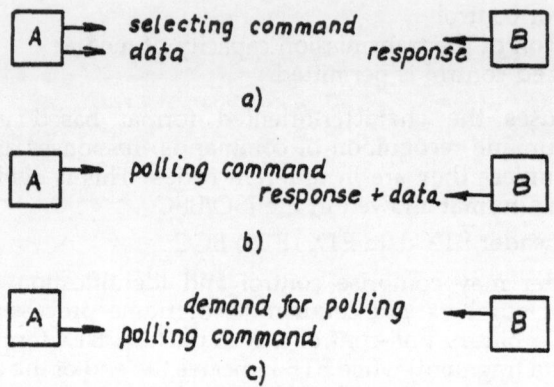

Fig. 4.29. ISO/BSC protocol modes. **a** Selecting; **b** polling; **c** contention.

Polling is supposed to be performed in an environment of relatively high data traffic intensity and short messages, which is a typical example of heavy conversational traffic. Otherwise the control station polls in vain and uses much of the time needed for other tributary stations. To avoid this the control station does not ask the tributary station but waits for its demand. The polling is substituted by a *contention* of tributary stations which, when ready to transmit data, require the polling command from the control station (Fig. 4.29c). The control station must provide certain pauses between communication to give the tributary station space for its demands. It is an example of centralized control when the initiative is given to tributary stations but the control station retains the managing function.

An example of data exchange between two stations controlled by the ISO/BSC protocol is given in Table 4.12. The control station A opens the communication with the polling command ENQ and with address B (step 1).

Table 4.12. Example of data interchange between a control station and a tributary station in the ISO/BSC protocol

Step	Station A (control)	Station B (tributary)
1	B, ENQ	
2		STX, data, ETB, BCC
3	ACK	
4		STX, data, ETX, BCC
5	NAK	
6		STX, data, ETX, BCC
7	ACK	
8		EOT
9	ENQ	
10		ACK
11	STX, data, ETX, BCC	
12		No response
13	ENQ	
14		ACK
15	EOT	

If station B has data to transmit, it reacts by sending the first frame carrying the first fragment of data terminated by ETB (step 2). After checking the BCC, station A confirms the frame with ACK (no errors have been detected) which serves as a demand for the following frame (step 3). The second data fragment is also the last one; it is terminated by ETX, not ETB (step 4). Suppose an error occurs during the transmission of the second frame. If it is detected, station A asks for retransmission by means of NAK (step 5). Station B retransmits (step 6), station A acknowledges (step 7), and station B announces the termination of transmission (step 8). The transmission, however, can be terminated only by the control station, but the latter may continue, if it has data to send. This case is shown in the table. Station A selects station B again by ENQ (step 9), etc. It may happen that after a command or after a frame no reply (neither ACK nor NAK) appears during a time-out (see Section 4.2.5). If the time-out expires without any reaction from the opposite station (step 12) a recovery procedure starts, for example by the ENQ request (step 13). The ultimate end of the communication is assessed only by the control station (step 15).

Even though the ISO/BSC link protocol is sufficiently sophisticated, has been used in many DTEs for a long time, and covers almost all applications required, it is less efficient than the more sophisticated bit-oriented protocols, HDLC in particular. However, it has survived because of its simplicity, particularly in cheap start–stop terminals. Therefore, there have been efforts to match the two types of protocols to each other. This is discussed in Section 4.6.3.

4.6.3 Bit-Oriented HDLC Protocols [10,14,17,23,33,42,45,47,52,58]

All bit-oriented protocols employ frames which are formed by strings of bits, that is, their length is not, in general, an integral multiple of characters or octets. The structure of the HDLC frame is shown in Fig. 4.30. It consists of several fields: flag, address field, control field, information field (only in frames carrying user data and in unnumbered frames informing about the reason of frame rejection), error control field, flag. The flag is the octet 01111110 which is designated for delimiting frames and for phasing (distinguishing beginnings and ends of frames). In order to avoid confusion between the real flag and a flag simulation within the frame (the information field of the flag being code-transparent may involve any combination of bits, so that it also enables the occurrence of the combination 0111110 which would be interpreted as the closing flag), a "0" is inserted after five consecutive "1"s at the transmitting side and deleted at the receiving side. The stuffing of bits, however, changes the size of fields and means the frame is no longer an integral multiple of eight and that the protocol is not suitable for parallel transmission by octets.

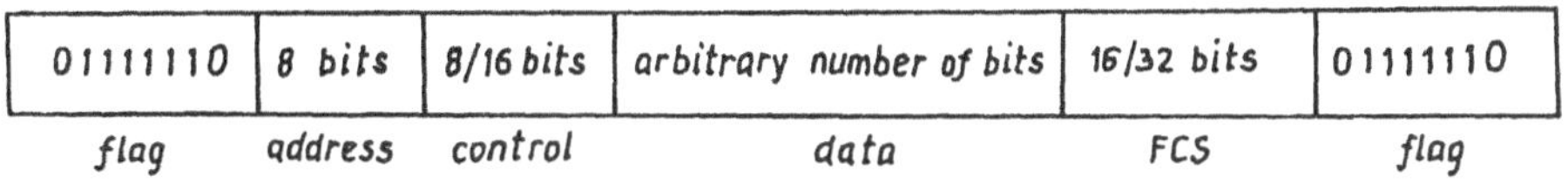

01111110	8 bits	8/16 bits	arbitrary number of bits	16/32 bits	01111110
flag	address	control	data	FCS	flag

Fig. 4.30. HDLC frame structure.

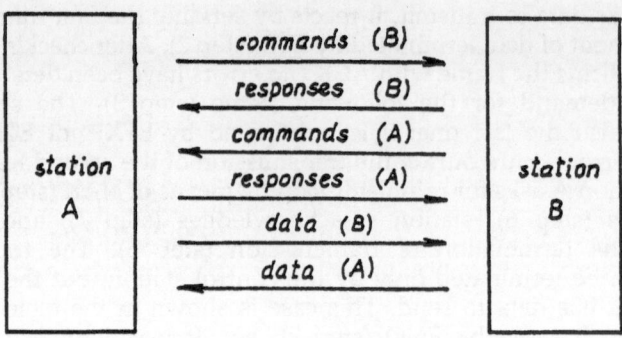

Fig. 4.31. Addressing policy in PSPDNs within the link layer.

The address field is used for the identification of stations. A command frame (frame carrying a command) contains the address of the destination station while a response frame contains the address of the station sending the frame (see Fig. 4.31). (For coding the address in PDNs see Table 4.16.) Normally, the address field consists of one octet for addressing up to 256 stations. If necessary, this field can be extended to two or more octets.

The control field is composed of one or two octets (basic or extended operation, respectively). It is divided into several subfields varying with the type of frame (information I, supervisory S and unnumbered U) and with the mode of operation (basic, extended). Its structure is shown in Table 4.13, the values of S bits are given in Table 4.14 and those of M bits in Table 4.15. The lower order bit distinguishes information frames from control frames; control frames are further differentiated by the second bit. The receive sequence number $N(R)$ indicates that the station has received without detecting an error all I-frames numbered up to and including $N(R) - 1$ and is expecting the

Table 4.13.   HDLC control field formats

Type of format and operation	Control field bits															
	1	2	3	4	5	6	7	8	9	10	11	12	13	14	15	16
I, basic (modulo 8)	0	$N(S)$			P	$N(R)$										
S, basic (modulo 8)	1	0	S	S	P/F	$N(R)$										
U, basic/extended (modulo 8/128)	1	1	M	M	P/F	M	M	M								
I, extended (modulo 128)	0	$N(S)$							P	$N(R)$						
S, extended (modulo 128)	1	0	S	S	X	X	X	X	P/F	$N(R)$						

Table 4.14. HDLC supervisory commands and responses

Designation	Significance	S bit combination 3	4	Application
RR	Receive ready	0	0	LAPB, LAP, LAPD
REJ	Reject	0	1	LAPB, LAP, LAPD
RNR	Receive not ready	1	0	LAPB, LAP, LAPD
SREJ	Selective reject	1	1	LAPB, LAP, LAPD

Table 4.15. HDLC unnumbered commands (C) and responses (R)

C/R	Designation	Significance	M bit combination 3	4	6	7	8	Application
C/R	UI	Unnumbered information	0	0	0	0	0	LAP, LAPB, LAPD
C	SNRM	Set normal response mode	0	0	0	0	1	
C	DISC	Disconnect	0	0	0	1	0	LAP, LAPB, LAPD
R	RD	Request disconnect	0	0	0	1	0	
C	UP	Unnumbered poll	0	0	1	0	0	
R	UA	Unnumbered acknowledgement	0	0	1	1	0	LAP, LAPB, LAPD
C/R	TEST	Test	0	0	1	1	1	
C	SIM	Set initialization mode	1	0	0	0	0	
R	RIM	Request initialization mode	1	0	0	0	0	
C	FRMR	Frame reject	1	0	0	0	1	LAPB, LAPD
R	CMDR	Command reject	1	0	0	0	1	LAP
C	SARM	Set asynchronous response mode	1	1	0	0	0	LAP
R	DM	Disconnect mode	1	1	0	0	0	LAPB, LAPD
C	RSET	Reset	1	1	0	0	1	
C	SARME	Set asynchronous response mode extended	1	1	0	1	0	
C	SNRME	Set normal response mode extended	1	1	0	1	1	
C	SABM	Set asynchronous balanced mode	1	1	1	0	0	LAPB
C/R	XID	Exchange identification	1	1	1	0	1	LAPD
C	SABME	Set asynchronous balanced mode extended	1	1	1	1	0	LAPB, LAPD

I-frame numbered N(R). The send sequence number N(S) denotes the sequence number of the next I-frame to be transmitted. Both numbers cycle from 0 through modulus minus one where the modulus equals either 8 (basic operation) or 128 (extended operation). The numbering serves for error control and flow control by the window mechanism (the maximum window size is 7 or 127 frames).

Finally, the P/F bit is intended for polling a response (poll bit P) in command frames and for indicating a response frame (final bit F). It may also serve as acknowledgement for error control. The error control field, called in HDLC "frame check sequence"(FCS) field, is based on the property of polynomial error-detecting codes. Since for the FCS field two octets are reserved, the code is generated by the polynomial $X^{16} + X^{12} + X^5 + 1$. The 16 bits of FCS represent the remainder of dividing by this polynomial the product of X^{16} multiplied by the content (represented as a polynomial) of the frame between two consecutive flags excluding the bits of the flags and all "0"s inserted for transparency. At the receiving side a decoding procedure similar to the encoding procedure is applied and if no errors are detected, the remainder would give a sequence of 16 "0"s. For certain control purposes (synchronization, phasing) slight modifications in computing the remainder are used, which results in the unique non-zero remainder in the absence of errors; this is 0001110100001111. All other patterns indicate the presence of detected errors and require retransmission.

Even though the chosen error-detecting code combined with the ARQ continuous (with the REJ command) or selective repeat (SREJ command) method is adequate for real circuits (the entire error-control mechanism decreases the bit error rate of 10^{-4} in a data circuit by six orders to 10^{-10} for frames of length 1000 bits) some applications would require an even lower residual error rate. The HDLC error control allows doubling of the FCS field applying a code generated by a polynomial of the degree of 32:

$$X^{32} + X^{26} + X^{23} + X^{22} + X^{16} + X^{12} + X^{11} + X^{10} + X^8 + X^7 + X^5 + X^4 + X^2 + X + 1$$

which is applied in several LANs.

Note that the coding scheme guarantees error protection not only of user data but also of all commands, responses and identifiers. Moreover, the bit-oriented structure of frames is code-independent, achieved by a very simple technical solution.

The HDLC principles enable us to create a set of protocols based upon several types of formats (Table 4.13), choice of commands and responses from Tables 4.14 and 4.15, three types of stations (primary – control, secondary tributary, combined – control/tributary), three basic modes (normal response mode, asynchronous response mode, asynchronous balanced mode) and three auxiliary modes (normal disconnect mode, asynchronous disconnect mode, initialization mode). For the time being, three classes of HDLC protocols are defined: unbalanced normal class, unbalanced asynchronous class, and balanced asynchronous class, with 14 options which meet practically all requirements and applications.

The unbalanced normal class is intended for centralized control of a communication over point-to-point as well as multipoint data circuits where the primary station is responsible for the opening and closing of a transmission, control of data flow in both directions (by means of the method of rejection of the RNR command and permission by the RR command or by the window mechanism agreed for a certain period of time) and solving exceptional events if any. For point-to-point data circuits the unbalanced asynchronous class of protocols can be used. These differ from the former in the degree of freedom of a secondary station: the former case is strictly centralized similarly to polling and selecting when the secondary station is allowed to transmit only upon an explicit permission from the primary station (cf. Fig. 4.29a, b) while in the latter case a secondary station may demand transmission similarly as in contention (Fig. 4.29c).

Fig. 4.32. HDLC balanced asynchronous mode.

The balanced asynchronous class of protocols is designed for decentralized control of communication over point-to-point data circuits (Fig. 4.32). The two combined stations have the same rights and duties, but thanks to asynchronous operations the control is modest in the exchange of overhead commands and responses. This is a reason for applying this class in PDNs.

The so-called *link access procedure balanced* (LAPB), which is the basic link protocol for access of packet-mode DTEs to PSPDNs, belongs in fact to the balanced asynchronous class with the option of REJ for higher efficiency and that of command I-frames. However, it does not allow the selective repeat method for error control. If necessary, LAPB is enhanced by extending the control field (modulo 128). For existing PDNs and DTEs the unbalanced asynchronous mode is permitted, and the link access procedure (LAP), with the same options as LAPB except extension, is temporarily allowed. However, the preference in new implementations is given to the LAPB protocol.

To improve reliability and efficiency the multiple physical circuit is anticipated. Beside the single link procedure (SLP) the multilink procedure (MLP) is defined. It allows the network layer to distribute and allocate data messages among several independent physical connections between stations. The SLP controlling separate data links can be arbitrary (HDLC, ISO/BSC), and the MLP performs as an intelligent arbiter and acts as an interface between the link and the network layer. If the HDLC (say LAPB) is used as the SLP, the control field of multilink format is hidden in the information field of the HDLC format.

The multilink procedure is recognized directly in the address field of LAP or LAPB. The coding of address fields is given in Table 4.16 where a DTE and a DCE are assigned to stations A and B of Fig. 4.32, respectively. For communication between signalling terminals at the boundary of different PSPDNs or between stations inside a PSPDN the assignment is arbitrary, while for telematic services the calling station and called station are recommended to be station A and station B, respectively.

For identification purposes, as in data transmission over switched access circuits (over a PSTN for example), the XID command/response is of use, and this option can be applied to LAPB. Moreover, data transmission over the

Table 4.16. Coding of address field in the LAPB single link and multilink protocol

Station	Code	
	Single link	Multilink
A-DTE	1 1 0 0 0 0 0 0	1 1 1 1 0 0 0 0
B-DCE	1 0 0 0 0 0 0 0	1 1 1 0 0 0 0 0

PSTN requires half-duplex transmission, which is not admitted in LAP and LAPB. To manage the direction of data flow a special module – the half-duplex transmission module – has been introduced, which looks for more than 15 consecutive "1"s in the bit sequence and checks the carrier for signalling a request for change of transmission direction. The LAPB equipped with this module has been named LAPX. It is suitable both for data transmission and some telematic services (teletex for example) whenever half-duplex media are available.

For the new transmission and switching support an ISDN also calls for an appropriate link protocol. As HDLC (in particular LAPB) has proved to be vital in principle, a derived protocol was proposed. It is intended for the packet data access over the D-channel (hence its name LAPD). The LAPD protocol differs slightly from the LAPB in frame structure (extension of the address field to two octets) as well as in certain procedures (single frame acknowledgement).

The HDLC, in fact, involves the connection establishment/release function and belongs to connection-oriented protocols. However, it is possible to derive a connectionless protocol from it using only unnumbered frames. For example, command/response UI, XID and TEST guarantee the delivery of data fragments, of course without acknowledgements and retransmissions. Such a protocol is suitable for LANs if connections within the link layer are not required. A comparison of the applications of the HDLC protocols discussed above (including the ISO/BSC protocol) is given in Table 4.17.

Returning to protocols for start–stop devices, we shall show how to bring the ISO/BSC and HDLC protocols together. Reconciling the two formats is surprisingly simple. The HDLC flag is replaced by the SYN character from

Table 4.17. Comparison of standardized link protocols

Application	Protocols			
	ISO/BSC	HDLC	LAP, LAPB, LAPD	LAPX
Control:				
Centralized	Y	Y	N	N
Decentralized	Y	Y	Y	Y
Communication:				
Two-way alternate	Y	Y	Y	Y
Two-way simultaneous	N	Y	Y	N
Circuit:				
Point-to-point	Y	Y	Y	Y
Multipoint	Y	Y	N	N
Transmission:				
Half-duplex	Y	Y	N	Y
Full-duplex	N	Y	Y	N
Serial	Y	Y	Y	Y
Parallel	Y	N	N	N
Synchronous	Y	Y	Y	Y
Start–stop	Y	Y	N	N

Y = yes; N = no.

IA5. Code transparency is preserved by inserting DLE ahead of each SYN within an information frame. Seven-bit characters are transformed into octets by adding one parity bit and the BCC character is extended to two octets which are generated by a polynomial of degree 16 as in the HDLC format. Both information frames and control (not carrying data) frames will be protected by the new pair of BCCs. The final format is:

SYN DLE STX address control (data) DLE ETX BCC BCC SYN

Moreover, the opening flag is not needed because synchronization is ensured by start and stop bits, and only one flag is sufficient for separating two closely consecutive frames (such a simplification is also allowed in the HDLC protocols).

The full explanation of at least one protocol out of the HDLC set is complicated because of the variety of situations and events which are covered by it. In addition, the survey of applications of HDLC commands/responses is space consuming and would hardly give the entire insight into the protocol syntax. We conclude this section as with the previous one: with a "snapshot" of a part of the DCE and DTE activity controlled by the widely used LAPB protocol (Table 4.18).

Suppose a DTE wants to transmit data toward its DCE representing the PDN. It addresses the DCE and sends the unnumbered SABM to place DCE in the asynchronous balanced mode non-extended (all control fields will be one octet in length) with the P bit set to one (command frame). In Table 4.18

Table 4.18. An example of data interchange between the DTE and DCE in the HDLC asynchronous balanced mode (LAPB)

Step	Station A (DTE)	Station B (DCE)
1	B, SABM, P=1	
2		B, UA, F=1
3	B, N(S)=0, N(R)=0, data	A, N(S)=0, N(R)=0, data
4	B, N(S)=1, N(R)=0, data	A, N(S)=1, P=1, N(R)=0, data
5	B, N(S)=2, N(R)=1, data	
6	A, RR, F=1, N(R)=2	B, RR, N(R)=2
7	B, N(S)=3, N(R)=2, data	B, RR, N(R)=3
8	B, N(S)=4, N(R)=2, data	B, RR, N(R)=4
9	B, N(S)=5, N(R)=2, data	A, N(S)=2, N(R)=5, data
10	B, N(S)=6, P=1, N(R)=2, data	A, N(S)=3, N(R)=6
11	A, REJ, N(R)=2	A, N(S)=4, N(R)=7, data
12	B, N(S)=7, N(R)=2, data	B, RR, F=1, N(R)=7
13	B, N(S)=0, N(R)=2, data	A, N(S)=2, N(R)=0, data
14	B, N(S)=1, N(R)=2, data	A, N(S)=3, N(R)=0, data
15	B, N(S)=2, N(R)=3, data	A, N(S)=4, N(R)=1, data
16	B, N(S)=3, N(R)=4, data	A, N(S)=5, N(R)=2, data
17	B, RNR, P=1, N(R)=5	
18		B, RR, F=1, N(R)=4
19	B, DISC, P=1	
20		B, UA, F=1

this step is briefly described as B, SABM, P = 1, omitting the flags and the error control field. In fact the following bit string is sent:

01111110	10000000	1111(0)100	11010111 11(0)111011	01111110
flag	DCE	SABM	frame check sequence	flag
	address	with P = 1		

where (0) represents bits inserted for transparency. The command may be refused by the DM response if the called DCE is in the disconnected state and cannot initiate the set mode command or accept it. Suppose the latter case and the DCE acknowledges the mode: it replies with its address, response UA and the F-bit set to one as a result of the soliciting command. The real bit string is of course:

0111110	10000000	11001110	11000001 11101010	01111110
flag	DCE	UA	frame check sequence	flag
	address	with F = 1		

Note that five consecutive "1"s did not appear between the flags and hence no insertion was needed. From now on the connection is established and the stations can transmit data in I-frames.

The two stations number independently their I-frames modulo 8, that is, by $N(S) = 0,1,2,...,7,0,...$ and at the same time they inspect the numbers of received I-frames. If a station receives an I-frame with $N(S) = i$ without detecting an error, it increments the number by one and replies by $N(R) = i + 1$, standing for the number of the expected I-frame. For example, in step 3 the DTE sends its first frame with $N(S) = 0$ and expects the first frame $(N(R) = 0)$ from the DCE. All I-frames in Table 4.18 are designated as data. Note that contrary to Table 4.12 the transmission is full-duplex and each response to a command or data follows two steps later.

The acknowledgement of the correct receipt of the I-frame numbered $N(S)$ is done by setting $N(R)$ to $N(S) + 1$ in the next transmitted I-frame (if both stations transmit data) or by the RR frame with $N(R) = N(S) + 1$ indicating that all I-frames numbered up to and including $N(S)$ have been acknowledged (steps 6 to 8). The transmitting station, however, may ask for acknowledgement with the P-bit set to one (step 4) which has to be followed from the opposite side by the RR-frame with F = 1 (step 6). On the other hand, as far as the numbering in modulo 8 or 128 is concerned, several (at most 7 or 127 frames, respectively) may be transmitted without explicit acknowledgement (window size).

Suppose an erroneous frame is detected. This may be caused by noise during transmission (false FCS), by an $N(S)$ sequence failure, by an invalid frame structure etc. (DCE I-frame with $N(S) = 2$ in step 9). After detecting it at the DTE side (step 10) the DTE sends the REJ-frame which refuses all DCE frames beginning with $N(S) = 2$ (step 11). The DCE transmits the frames numbered 2, 3 and 4 in steps 13, 14 and 15, respectively.

As soon as any station is not able or does not want to continue to receive data, it may interrupt the process by the RNR-frame (see the DTE in step 17), acknowledging at the same time only the last frame correctly received (DCE I-frame with $N(S) = 4$). The following frames remain unacknowledged (DCE I-frame 5). The DCE acknowledges all DTE frames numbered 3 or less (step 18). The DISC unnumbered command sent by any station is used to terminate the mode previously set. In our example it is sent from the DTE side (step 19) after which the DCE reacts with the UA response (the final step, step 20).

Besides commands and responses, field structure and contents (N(R), N(S), P/F-bit), and corresponding procedures together with several time-outs, window sizes and moduli of numbering provide for the detection of exceptional states and their recovery. These protocol parameters may assume different values at different stations (at the DTE and DCE side for example) but these have to be agreed for a period of time and made known to each of them.

4.7 Network Layer

4.7.1 General Considerations [14,17]

The network layer is important for the establishment of circuits in a switched environment (PSTN, PDN, ISDN), for interworking and for implementing higher user layers. As two types of switching (circuit and packet) are the commonest in use, it is expected that different mechanisms will be applied to provide the necessary functions and services. There are significant differences between protocols controlling the execution of network functions. Circuit switching protocols model their operation on the PSTN, which has survived and supported most telephone services as well as some data transmission services. For most types of data transmission, however, the low quality of service supported by the PSTN, including delays caused by call establishment/release, less automation in the course of procedures, and limited user facilities, have highlighted the need for the development of new network protocols. Nevertheless, as we shall see later, the network protocol for CSPDNs took over most of the classical telephone procedures.

Packet switching network protocols are based on the exchange of packets – units of sizes, structures and contents determined in advance. This may be time-consuming, but is compensated by providing enough space for all the necessary information which should be exchanged during the control phase as well as in the data transfer phase. Packets also search a pass in a hop-by-hop manner and need co-operation of all network elements which have already been activated.

Two modes are allowed in the network layer: connection-oriented and connectionless (datagrams). The latter is, of course, provided in the packet switched environment but it may support the control of circuit switching provided that this control is performed via separate channels as, for example, in the case of the ISDN.

The basic principles of circuit switched and packet switched protocols are very similar, but differ in protocol units, their structures and contents and the actions required by different services and user facilities. The decisive divergences stem from the mode. We can demonstrate this by examining the handshaking of subscribers and a network for four typical cases (Table 4.19).

Let us first examine the upper mode. A subscriber demands a service from the network (to establish an appropriate connection with the called remote counterpart, for example). The network may reject the demand because there are congestions, derangements, faults in demand, or unavailability of the

Table 4.19. Principle of network control

Mode	Action	Subscriber	Network	Subscriber
Connection oriented	Unsuccessful	Service demand ⟶		
			⟵ Demand rejected	
	Successful	Service demand ⟶		
			⟵ Demand accepted	
			Demand delivery ⟶	
				⟵ Demand rejected/ accepted
			⟵ Reply delivery	
Connectionless	Unsuccessful	Service demand ⟶		
			⟵ Demand rejected	
	Successful	Service demand ⟶		
			⟵ Demand accepted	
			Demand delivery ⟶	
				Demand rejected/ accepted

called subscriber, and informs the calling subscriber (unsuccessful demand). If, on the other hand, the demand is accepted, the network delivers it to the addressee. The addressee is either not ready to accept the demand and rejects due to, say, other activities, or it agrees and informs the network (successful demand). The reply of the called subscriber is finally delivered to his calling counterpart. The amount of information acquired during connection establishment strongly increases in CSPDNs and PSPDNs as compared with the PSTN (see the following sections.)

Such handshakings are valid if the network provides the connection-mode service when switched (physical or virtual) circuits are established, maintained and released by means of routing, relaying and multiplexing.

The connectionless-mode does not provide a connection and the calling subscriber only passes the demand to the network for delivery to the addressee. The network may or may not inform the calling subscriber of its state but in most cases the calling subscriber does not know how his demand will be treated. Even though the connectionless-mode is simpler to realize, PDNs usually do not provide it. However, it is very often applied in LANs and private data networks.

The appearance and subsequent approval of the OSI reference model brought problems because there were many PDNs which could not follow hitherto non-existing protocols. The OSI protocols affected private networks (computer, terminal, local) due to their relatively short lifetime. Capital expenditures for the development of PDNs do not allow frequent changes, whereas local and middle area user networks can be modified and replaced more often.

As the network layer of most PDNs is not capable of supporting standard OSI network services, the openness of these networks may fail and PDN users could be limited in their demand. Therefore the structure of the network layer had to be modified accordingly. ISO 8648 defines three network sublayers hierarchically arranged:

• The lowest network access sublayer and protocol
• The convergence sublayer and network dependent protocol
• The highest convergence sublayer and network independent protocol

The corresponding sublayer protocols execute functions to provide services to higher sublayers up to the OSI reference model transport layer. The access protocol is strongly network-oriented: the X.21 protocol (see Section 4.7.2) and X.25 network level protocol (see Section 4.7.3) belong to this group. The complexity of the middle sublayer protocol depends on how much the network access sublayer differs from the standardized OSI network layer. For example, if the lowest sublayer limits the length of data strings, the convergence protocol should perform the segmenting/reassembling function. Finally, the highest network sublayer is necessary if the network consists of several networks in tandem, with each network providing different services. This sublayer and the corresponding protocols support the idea of harmonizing separate networks at the network layer by smoothing out all divergencies. Two types of network control (and corresponding protocols) are recognized in public networks:

• User-to-network control or, briefly, access control (protocol)
• Intranetwork control (protocol)

Public network operating organizations (mostly PTT administrations) strictly guard accesses to protect the network from unauthorized and detrimental infringement by users. This is the reason for promoting international standardization of access protocols. On the other hand, intranetwork control is a matter of national decision. This applies to the whole network architecture, even though the physical layer and link layer access protocols are often spread throughout the network. A different approach has been chosen for network layer protocols regardless of the type of switching. Access protocols at the DTE–DCE interface may require control information different from control information for operations within a network.

The development of the ISDN further made its mark on network control by separating the control path from the user information path. The in-band or in-slot control signalling typical for all PDNs has given way to out-slot or out-of-band control signalling over separate physical or virtual channels. If out-slot control signalling simultaneously supports user data flows, it is called common channel control signalling. This is used in ISDN intranetwork control (Section 4.7.3).

To conclude, a terminological note might be of help in understanding the continuity of development from old to new. While according to the OSI concept communication is governed by protocols, in the circuit switched environment the term "signalling" as protocol action has been used and survives even in the OSI era. In order not to confuse the reader we shall in future respect historical development by giving preference to the term "signalling" in dealing with circuit switching in Section 4.7.2, and to the term "protocol action" in Section 4.7.3 on packet switching. After all, signalling and protocol actions are only two sides of one coin.

4.7.2 *Circuit Switching Protocols* [33,47,72]

The network layer protocols of CSPDN are influenced by classical control methods in the PSTN. However, the standardization of user-to-network as well as intranetwork protocols anticipated the development of digital transmission means and CCITT first proposed the digital signalling interface X.21 for synchronous DTEs and X.20 for start–stop DTEs. The corresponding CCITT recommendations were approved as early as 1972. Practice, however, still called for control over analogue transmission by means of V-series modems so that four years later the corresponding twins (bis's) were added. The intranetwork signalling is either of the common-channel or channel-associated type. Common-channel signalling was standardized for CSPDNs applications as late as 1980 (when X.60 was first approved) while channel-associated signalling is the subject of much older recommendations: X.70 for asynchronous CSPDNs and X.71 for synchronous CSPDNs.

Let us focus our attention on the user-to-network or access signalling (protocol) on digital circuits. Table 4.20 shows an example of user access to the PSTN and will serve as a reference for the next table (Table 4.21), illustrating the main features of signal exchange at the DTE–DCE interface on the X.24 interchange circuits (Section 4.5).

The on-the-hook state of the telephone subscriber corresponds to the ready state of the CSPDN user (DTE). It is characterized by the signal conditions of all interchange circuits (T, R, C, I) being set to one (note that the continuous OFF condition in control circuits is the same as binary 1 in data circuits and continuous ON is the same as binary 0). The signal conditions on interchange circuits are designated in Table 4.21 by t, r, c and i. The lifting of the handset (off-the-hook) corresponds to the transition of signal conditions t and c of the calling DTE. When the local DCE (DCE A) is ready to accept a call, it begins to transmit a sequence of IA5 characters 2/11 (+) on the R circuit corresponding to the dialling tone in the PSTN with the meaning "proceed-to-select". The dialling in CSPDNs is accomplished by sending (over the T circuit) a series of IA5 characters with odd parity for synchronous DTEs and with even parity for start–stop DTEs, in other words a sequence of octets. They carry not only

Table 4.20.   An example of the user-to-PSTN protocol

Subscriber A (calling)		PSTN	Subscriber B (called)
On-the-hook		Idle	On-the-hook
Off-the-hook			
	←—— Dialling tone		
Dialling of number	—→		
	←—— Ringing tone		Ringing —→
			Off-the-hook
Conversation	←—→		←—→ Conversation
On-the-hook requested —→			
			←—— On-the-hook accepted

Table 4.21. An example of the user-to-CSPDN protocol (DTE–DCE interface)

DTE A (calling)	CSPDN		DTE B (called)
	DCE A	DCE B	
Ready (t=1, c=OFF)	Ready (r=1, i=OFF)		Ready (t=1, c=OFF)
Call request (t=0, c=ON) $\longrightarrow$			
	$\longleftarrow$ Proceed-to-select (r=+, i=OFF)		
Selection signals (t=IA5, c=ON) $\longrightarrow$			
	$\longleftarrow$ Call progress (r=IA5, i=OFF)	Incoming call (r-BELL, i=OFF) $\longrightarrow$	
			$\longleftarrow$ Call accepted (t=1, c=ON)
	$\longleftarrow$ Ready for data (r=1, i=ON)		$\longrightarrow$
Data exchange (t=data, c=ON)	$\longleftrightarrow$ (r=data, i=ON)	(r=data, i=ON) $\longleftrightarrow$	Data exchange (t=data, c=ON)
Clear request (t=0, c=OFF) $\longrightarrow$			
		Clear indication (r=0, i=OFF) $\longrightarrow$	
			$\longleftarrow$ Clear confirmation (t=0, c=OFF)
	$\longleftarrow$ Clear confirmation (r=0, i=OFF)		

addresses but also facility request, facility registration and/or facility cancellation codes.

While most PSTNs are able to inform the calling subscriber only about the status of the network (free or busy) and of the called subscriber (free, busy or not obtainable), the CSPDN is more user-friendly and offers more comprehensive information in the form of call-progress signals, described in Table 4.22. These signals support fully automatic operation evoking the appropriate DTE responses (see last column of Table 4.22). The incoming call in a CSPDN (a sequence of IA5 characters 0/7 transmitted by the DCE of the called subscriber) corresponds to ringing in the PSTN. However, if in the CSPDN the called DTE accepts the call (corresponding to answering a call in the PSTN), the DCE of the called subscriber may proceed by sending IA5 characters which provide information about charging, subaddressing, user facilities, call characteristics, etc. (not shown in Table 4.21).

Now a full-duplex connection has been established and either DTE can send data under the control of any link protocol (see Section 4.4). When all the data has been exchanged either DTE can request to clear the connection by setting the signal condition c to OFF. However, it must stay in the state of readiness to receive data as long as it does not get the clear confirmation

Table 4.22.   Call progress signals used in CSPDNs

Signal	Interpretation	Recommended action of calling DTE
Terminal called	The call has been sent and is being awaited for acceptance	Waiting
Redirected call	The call has been redirected to another address indicated by the called DTE	
Connect when free	The called DTE is occupied and the call is in queue until a line becomes free	
No connection	The connection has not been set up, no specified reason being indicated	Retry
Number busy	The called DTE is in connection with another DTE and is not ready to accept the incoming call	
Transmission error of selection signals	An error in the selection signals detected by the adjacent switching centre	
Network congestion	Short-term congestion due to errors in internal protocols, for example	
Access barred	Unauthorized access or incompatible closed user group	Retry later (an immediate attempt would result in the same signal)
Changed number	The called DTE has been assigned a new number	
Invalid facility request	A facility requested by the calling DTE is detected as invalid (not subscribed, not available) by the local DCE	
Incompatible user class of service	The called DTE is incompatible with the user class of service of the calling DTE	
Not obtainable	The called DTE number is not assigned to any DTE	
Call information centre	For details of the called number which is temporarily unobtainable	
Controlled not ready	Refers to the called DTE	
Long-term network congestion	A major shortage of network resources	
Local procedure error	A protocol error caused by the calling DTE is detected by the local DCE	
Network fault in local loop	Refers to the local loop associated with the called DTE	
DCE power off	Refers to the called DCE	
Uncontrolled not ready	Refers to the called DTE	
RPOA out of order	The breakdown of the RPOA nominated by the calling DTE	

signal from the remote DTE. In our case DTE A starts clearing the connection and DTE B responds to DCE B's clear indication by the clear confirmation signal. Finally DTE A sets t to one as an acknowledgement of the clear confirmation signal from the network and both DTEs as well as both DCEs return to their ready states awaiting another call.

The protocol is, of course, more complicated, because it must solve certain exceptional situations and perform maintenance by means of test loops. However, one situation – a call collision – may occur rather often and is therefore controlled by the protocol. This situation occurs when an incoming call and an outgoing call appear simultaneously. According to the rule that outgoing calls are always preferred, because of problems with the reallocation of DTE resources committed to the outgoing call, the incoming call is cancelled.

CSPDN user facilities are explained in detail in Chapter 5 and it should be mentioned here that all these facilities are supported by the protocol. For example, when the direct call facility is assigned for an agreed contractual period, selection signals are always bypassed.

It should be noted that in the case of synchronous DTEs each sequence of IA5 characters is preceded by at least two SYN (IA5 1/6) characters. This is dropped in the case of start–stop DTEs where the formation of start–stop signal characters includes the addition of the even parity bit.

A complete description of CSPDN protocols by means of state diagrams, accompanied by examples in the form of sequences of events, is to be found in CCITT recommendations or reprinted in handbooks and other publications. The state diagrams cover all states but they cannot supersede time sequences. Sequences of events, on the other hand, are furnished with values of time-outs needed for the recovery from exceptional states but they present illustrative examples rather than full protocol specifications.

If digital circuits are not available for accessing the CSPDN, analogue circuits (based on telephone channels or 48 kHz primary groups) equipped with corresponding V-series modems have to be applied. In that case the X.24 interchange circuits are substituted by the V.24 interchange circuits. This is possible because a mapping of X.20/X.21 signals protocol signals onto states of selected V.24 circuits is standardized (Table 4.23). The V.24 interface,

Table 4.23. Relation between protocol elements for digital (X.20/X.21) and analogue (X.20bis/X.21bis) accesses to CSPDNs

Protocol signals	States of V.24 interchange circuits	
Call request	108.1	ON
Incoming call	125	ON
Call accepted	108[a]	ON
Ready for data	107	ON
Clear request	108[a]	OFF
Clear indication	107	OFF
Clear confirmation	108[a]	OFF

[a] Circuit 108 means 108.1 or 108.2.

however, limits the PSPDN services and facilities because of lack of line identifications and call progress signals. Some disadvantages can be avoided by using the 200-series V.24 interchange circuits according to CCITT V.25.

It should be noted that the X.20/X.21 protocols do not provide a complete OSI network service. This is due to their early arrival on the scene (they can be placed among the already mentioned pre-OSI protocols). After some enhancement and mapping of protocol signals onto OSI network protocol data units and service data units it will be possible to achieve the provision of standard OSI network services which would help in interworking with other OSI networks. The CSPDN architecture makes it impossible for the X.20/X.21 protocols to recognize the boundary between physical layer and network layer, though several attempts to solve this problem have been made and are recorded in the literature (as well as in Section 4.5). Because of the transparency of the link layer this is only a question of implementation.

The ISDN network protocol will now be dealt with. Though the ISDN is, in principle, a circuit switched network, the control of switching is achieved (and so is the overall communication control) in the separate D-channel in the packet mode. B-channels are only subjected to switching, and all their layers, with the exception of the physical layer, are transparent. The packet switched network protocols will be dealt with in Section 4.7.3 and for that reason the ISDN network protocol is described there.

4.7.3 Packet Switching Protocols [8,10,14,23,31,41,42,47,54,69,72]

The specification of network layer services and protocols in the packet switched environment began to develop by incorporating HDLC within CCITT recommendation X.25 in 1976. The packet level, as it was called in X.25, has experienced many changes, improvements and modifications during three study periods as PSPDNs were commissioned: the second version was approved in 1980 and added a new service (connectionless-mode or datagram) and further user facilities (fast select), four years later the datagram service was dropped but new protocol parameter values were supplied (registration codes, new packet size). The 1988 version introduced modifications for interworking with the ISDN.

The X.25 protocol is, in principle, an access protocol (at the DTE–DCE interface in the three lower layers), the intranetwork protocol X.75 was finished in 1978 as a provisional version, formally approved in 1980 and subsequently amended on several occasions. However, ISO has not been idle. It proposed the network service definition in 1983 and approved it four years later as standard 8348, which later appeared as CCITT recommendation X.214. The joint work bore fruit by harmonizing the X.25 protocol with the OSI reference model, culminating in ISO 8878 and CCITT X.223. On the other hand, the X.25/1984 network protocol was the basis for ISO standard 8208. Finally, ISO has dealt with the connectionless-mode network services and the corresponding protocols (ISO 8873). (For other relevant documents referring to the network layer see Table 4.1.)

The network layer of packet switched networks (private as well as public) uses packets of different size, structures and contents defined in advance for the exchange of control and user data. The procedure of call set-up and clearing in packet switched networks providing the connection-mode service

Table 4.24. Basic packet types, their applications and identifiers

Function	Packet type		Virtual circuit service		Packet identifier (coding of octet) bits 87604321
	From DCE to DTE	From DTE to DCE	Switched	Permanent	
Call set up and clearing	Incoming call	Call request	o		00001011
	Call connected	Call accepted	o		00001111
	Clear indication	Clear request	o		00010011
	Clear confirmation	Clear confirmation	o		00010111
Data transfer and interrupt	Data transfer	Data transfer	o	o	xxxxxxx0
	Interrupt	Interrupt	o	o	00100011
	Interrupt confirmation	Interrupt confirmation	o	o	00100111
Flow control and reset	Receive ready	Receive ready	o	o	xxx00001
	Receive not ready	Receive not ready	o	o	xxx00101
	Reset indication	Reset request	o	o	00011011
	Reset confirmation	Reset confirmation	o	o	00011111
Restart	Restart indication	Restart request	o	o	11111011
	Restart confirmation	Restart confirmation	o	o	11111111

is comparable with the procedure applied in circuit switched networks (see Section 4.7.2) since the control packets may carry the same information as the signalling sequences. For example, the call request packet sent by a calling DTE determines not only the called DTE but also the required user facilities to be provided by the network. On the other hand, the call connected or clear indication packet routed to the calling DTE by the network confirms the negotiated facilities and may announce the quality of the demanded service, or explain the reasons why the call could not be accepted, respectively.

In addition to the services mentioned, packet switched networks are capable of supporting the connectionless-mode service. In that case the protocol is reduced to a simple emission of individual packets containing a limited amount of user data together with the necessary control information (see Table 4.19).

Control in the network layer, particularly if the connection-mode service is provided, requires, however, more types of packets. In order to better acquaint the reader with network protocols, the following examples are taken from standardized network protocols used widely in PSPDNs. To avoid confusion each term prefixed by X.25 or X.75 will refer to CCITT X.25 or X.75 packet level protocols, respectively (in these recommendations the three lowest layers of the reference model are called levels).

Table 4.24 lays out the basic set of X.25 packet types. The packets are arranged according to the functions which they serve and to the two services provided by any X.25 PSPDN (switched or permanent virtual circuit). Note that the permanent virtual circuit needs neither set-up nor clearing packets since the connection is permanently established for an agreed period of time. The table also contains coding for the third packet octet which identifies the type of packet. The "x"in a coded octet stands for 0 or 1, depending upon an indication for further use.

The packet identifier is designated to distinguish up to 256 packet types. However, only a small portion of them is defined or studied. For example, flow control and reset packets in Table 4.24 refer to the numbering modulo 8, but numbering modulo 128 is also permitted and this fact is indicated by setting bits 6, 7 and 8 to zero. Optional X.25 packets are diagnostic packets by which the DCE conveys information about unrecoverable situations needing intervention by higher layers at the DTE side, registration packets indicating the status and changes of user facilities, and the reject packet for the retransmission facility in the ARQ continuous error control method.

Each packet has its prescribed format made up of a packet header and a data field (user data, indication data, registration data or diagnostic data). As it is advantageous to form octets the X.25 packet format is octet-oriented (it is an integer multiple of 8-tuples of bits), though in some cases only semi-octets (4 bits) provide comprehensive information. For this reason the packet structure is usually shown as in Fig. 4.33. It represents a binary sequence according to the rule that octet 1 and bit 1 of each octet are transmitted first.

The general format identifier distinguishes between the two numbering schemes (modulo 8 and modulo 128) by bits 5 and 6 (10 and 01, respectively) and contains two control bits: delivery confirmation bit 7 (D-bit) which set to 1 is used for end-to-end acknowledgement of delivery of data packets, and the qualifier bit 8 (Q-bit) set to 1 indicates that data in the packet will serve to control a PAD.

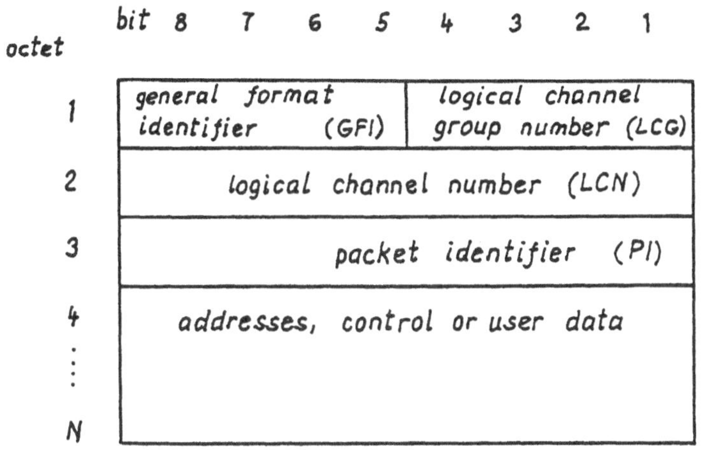

bit 8   7   6   5   4   3   2   1

octet

1	*general format identifier (GFI)* | *logical channel group number (LCG)*
2	*logical channel number (LCN)*
3	*packet identifier (PI)*
4 ⋮ N	*addresses, control or user data*

Fig. 4.33. The X.25 packet format.

The next 12 bits (one and a half octets) are reserved for logical channel numbering. The first semi-octet (bits 1 to 4) indicates up to 12 logical channel groups which are agreed with PTT administrations for a period of time. It is recommended that it begins with the group (or groups) used for permanent virtual circuits followed by groups of logical channels forming one-way incoming, two-way, and one-way outgoing virtual circuits.

The second octet carries the number of logical channels which, together with the logical channel group number, results in a capacity of 4096 independent channels. Note that this number is not a network address (see Section 4.2.4); it simply distinguishes multiplexed connections.

The third octet houses the packet identifier encoded in compliance with Table 4.24. Bit 1 is set to 1 in all packets, with the exception of data packets, which makes it possible to recognize control packets immediately. The remaining bits in data packets have the format defined by Fig. 4.34 for basic (a) and extended (b) formats. The latter requires, of course, two octets in the packet header. The M-bit (more data) indicates a sequence of more than one data packet bearing user fragments (or blocks from the transport layer): the

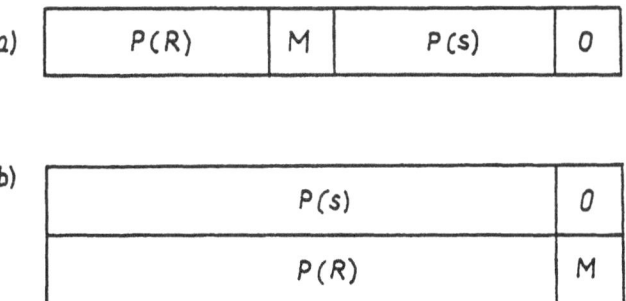

Fig. 4.34. Control octets of the X.25 packet header. **a** For basic (modulo 8) format; **b** for extended (modulo 128) format.

M-bit is set to 1 in all such packets except the last one, for which the M-bit is set to 0. P(R) and P(S) play the role of N(R) and N(S) of the HDLC frame structure: they indicate a packet receive sequence number and a packet send sequence number, respectively. The numbering is modulo 8 in the basic format or modulo 128 if the format is extended and serves flow control by means of the window mechanism.

The packet type identifier is followed by network addresses (in packets for call set-up and clearing for example), by coded status information (restart and reset packets), or by user data (data and interrupt packets). The address field contains the addresses of the called as well as the calling DTE. Each address is preceded by the address length mapped in binary decimal code to a semi-octet. This implies that the maximum length of each address is 15 decimal digits, which conforms with the PDN numbering plan of CCITT X.121 (Section 6.6).

In some control and indication packets (such as clear, reset, restart) the packet headers express the causes for clearing, resetting and restarting and possibly more explanatory information (diagnostics). Since an octet is assigned to each element of information, the set of causes and diagnostics is sufficiently rich (255 code elements). Even though only a small part of it is defined for the time being, the PSPDN user benefits from being well informed about unsuccessful calls, interruptions, closing, problems due to errors and faults of protocol, the remote DTE, and the network proper (national and international part).

The call progress signals are summarized in Table 4.25 (see Table 4.22 for CSPDNS) and are classified in two groups of possible DTE actions. The redundancy of the DTE waiting follows directly from the packet switching principle. In contrast to CSPDNs the space in packets permits the inclusion of more information so that, for example, the invalid facility request could be specified as an error in the facility or facility parameter value fields; the local procedure error can be more detailed still.

As an illustration of the X.25 network protocol a simple example of successful data exchange over a PSPDN, as seen on the DTE–DCE interface, is given in Table 4.26. Such an example cannot of course illustrate the whole protocol syntax, particularly the time relations. However, it opens the door into the field of PSPDN service providers.

All PSPDNs in operation more or less follow CCITT X.25, which specifies the user-to-network protocol. The network appears to the users as a black box seen only through the DTE–DCE interface. Table 4.26 respects this approach and shows only two communicating DTEs with their corresponding DCEs. In fact, the two DTEs communicate with each other over a switched virtual circuit.

All commands, responses and data are conveyed only within the network layer in packets. Hence DTE A, if it is the initiator of the communication, selects a free logical channel and sends the call request packet. This packet carries, besides the standard header, an indication of the user facilities to be negotiated. At DTE B the incoming call packet is transferred from DCE B which also chooses an appropriate (usually the first free) logical channel that need not be the same as for the DTE A–DCE A communication. DTE B can react in one of two ways upon receipt of this packet: if the call is accepted, the response is the call accepted packet; if the call is rejected, the clear request

Table 4.25. Call progress signals used in PSPDNs

Signal	Interpretation	Recommended action of calling DTE
Number busy	The called DTE is in connection with the other DTE and is not ready to accept the incoming call	Retry
Network congestion	Short-term congestion due to errors in internal protocols, for example	
Network out of order	Temporary inability to handle data traffic	
Network operational	The network is ready to resume normal operation after a temporary failure or congestion	
Remote DTE operational	The remote DTE–DCE interface is ready to resume normal operation after a temporary failure or congestion	
Access barred	Unauthorized access or incompatible closed user group	Retry later (an immediate attempt would result in the same signal)
Invalid facility request	A facility requested by any DTE is detected as invalid (not subscribed, not available) by the local DCE	
Reverse charging acceptance not subscribed	Refers to the called DTE	
Fast select acceptance not subscribed	Refers to the called DTE	
Incompatible destination	The remote DTE–DCE interface does not support a function or facility requested	
Not obtainable	The called DTE number is not assigned to any DTE	
Local procedure error	A protocol error caused by any DTE is detected by the local DCE	
Remote procedure error	A protocol error caused by any DTE or an invalid facility request by the remote DTE is detected by the remote DCE	
Out of order	The remote DTE is out of order due to its uncontrolled not ready, its DCE power off, a network fault in the local loop, or to the physical layer and/or link layer not functioning	
Network fault in the local loop	Refers to the local loop associated with the called DTE	
DCE power off	Refers to the called DCE	
Uncontrolled not ready	Refers to the called DTE	
DTE originated	The remote DTE has initiated a clear, reset or restart procedure	
RPOA out of order	The breakdown of the RPOA nominated by the calling DTE	

packet follows. Supposing the former case, the call connected packet appears at the DTE A–DCE A interface and announces the establishment of a switched virtual circuit. This opens the two-way simultaneous communication. All the necessary information concerning user facilities and values of quality-of-service parameters (throughput, end-to-end transit delay, see Section 7.1) has been negotiated during the call establishment phase and is specified in packets. In Table 4.26 these details are omitted and only packet types together with addresses are shown.

The two-way data flow is controlled by the window mechanism of the standard (default) size $W = 2$, that is, at most two consecutive packets can be sent without acknowledgement. In our example DTE A wants to transmit four packets numbered $P(S) = 0$ to $P(S) = 3$ but, because of the negotiated

Table 4.26. An example of data interchange over a virtual circuit at the PSPDN network layer

Step	DTE A	PSPDN		DTE B
		DCE A	DCE B	
1	Call request B, A →			
2			Incoming call, B, A →	
3				← Call accepted, A, B
4		← Call connected A, B		
5	Data A, P(S)=0 M=1, P(R)=0 →			
6	Data A, P(S)=1, M=1, P(R)=0 →		Data A, P(S)=0 M-1, P(R)=0	→ Data B, P(S)=0, ← M=1, P(R)=0
7		← Data B, P(S)=0 M=1, P(R)=0	Data A, P(S)=1 M=1, P(R)=0	→ Data B, P(S)=1, ← M=0, P(R)=1
8		← Data B, P(S)=1, M=0, P(R)=1		← Receive ready, P(R)=2
9		← Receive ready P(R)=2		
10	Data A, P(S)=2, M=1, P(R)=? →			
11	Data A, P(S)=3, M=0, P(R)=2 →		Data A, P(S)=2, M=1, P(R)=2 →	
12	Clear request B, A →		Data A, P(S)=3, P(R)=2 →	
13			Clear indication B, A →	
14				← Clear confirmation A, B
15		← Clear confirmation, A, B		

window size it is allowed to send only the first two packets unless it gets an acknowledgement from DCE A. DTE B also sends data, but only in two packets. To enable DTE A to continue transmission DTE B replies with the receive ready packet with P(R) = 2, which releases DTE A to send the remaining data packets.

After exchanging all data any party (either DTE or DCE) may initiate the clearing phase. As shown in Table 4.26, DTE A opens the clearing process by the clear request packet. To DTE B this request is indicated by the clear indication packet and DTE B responds by the clear confirmation packet. From that instant the virtual circuit is definitely released.

A similar procedure holds even in the case of data transfer over a permanent virtual circuit. As the circuit has already been established, only the middle phase of data exchange takes place. Moreover, negotiations about user facilities and quality of service are omitted because they have been dealt with previously.

Table 4.26 does not take into account exceptional states such as faults, errors, collisions, etc. The X.25 PSPDN treats most of these by introducing special packets and corresponding procedures. For example, interrupt packets solve interrupt procedures without influencing flow control (sequence numbering is preserved) and allow the exchange of up to 32 octets of expedited data. Reset packets are used to reinitialize a virtual circuit. If the data flow fails to recover after interruptions, the procedure for reset is applied. This procedure reinitializes each virtual circuit and removes all data packets and interrupt packets. Simultaneously, the window mechanism sets all numbering to zero. The reset packet contains an indication of the cause of reset, such as network congestion, protocol error, out of order, etc. To avoid collisions and to recover after infrequent events a set of recovery procedures based upon time-outs is employed.

The X.25 network protocol is designed for PSPDNs. Moreover, its harmonization with other standard network protocols allows its use for wider applications (see CCITT X.223 and ISO 8878). It is, however, described in terms of the DTE–DCE interface, as the access protocol. The internal arrangement of the network protocol within the network (intranetwork protocol) is left to the network provider. It may implement the X.75 protocol originally designed for interworking on the three lowest layers between PSPDNs. This interworking is delegated to the signalling terminals each terminating a PSPDN. By replacing a signalling terminal with a switching exchange the X.75 protocol is applicable as an intranetwork protocol as well.

An example is shown in Fig. 4.35 and does not need any particular comment. In contrast to Table 4.26, the figure shows the packet dissemination over a PSPDN: for example, the call request packet traces a virtual call by occupying logical channels between switching exchanges, the call connected packet is returned over the same logical channels after the last call request packet had appeared at a called DCE as the incoming call packet. The names of packets in parenthesis refer to the X.25 protocol.

Whatever intra- or internetwork protocol is chosen it must ensure that all DTE control and indication data, including types of service and user facilities, will be preserved in packets during their passage through the networks and conveyed to the other DTE. The X.75 protocol meets these requirements; moreover, its call set-up and clearing packets support internetwork services, if required.

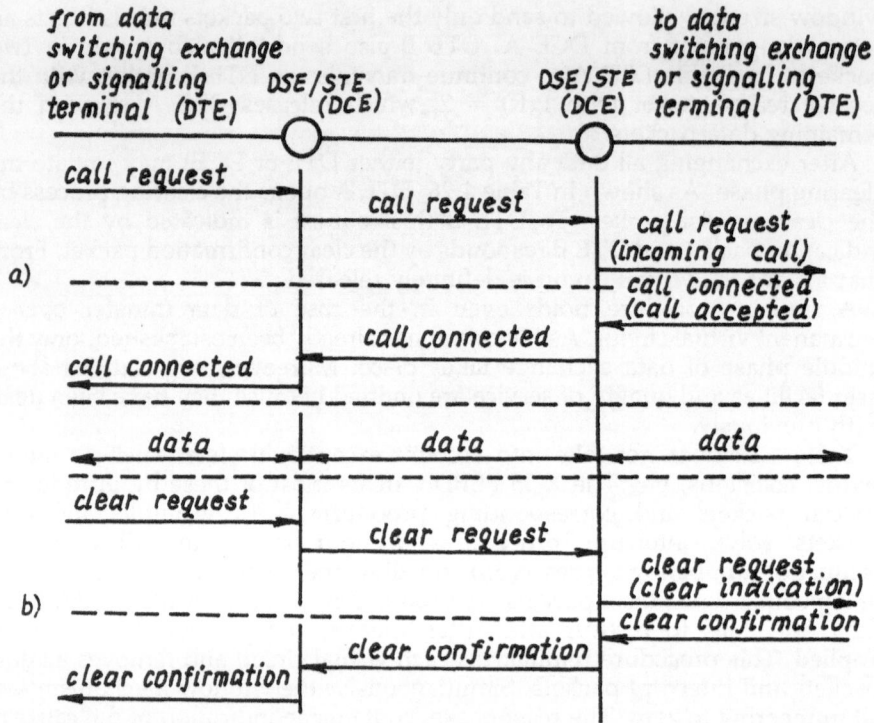

Fig. 4.35. An illustration of **a** call set-up and **b** clearing within a PSPDN or between PSPDNs.

Although the network layer and the network protocol need not guarantee the end-to-end confirmation (this is a task for the transport layer and its protocols), the X.25 protocol enables the acknowledgement of packet delivery over a network connection from the receiving DTE by means of the D-bit. Such a service is optionally included among network services within the OSI reference model just to support the X.25 protocol.

The X.25 PSPDNs do not provide the connectionless-mode service because this was abandoned during the CCITT study period 1980– 1984. It has been superseded by the "fast select" user facility, described below.

The fast select procedure consists of the following: if the DTE has subscribed to the fast select acceptance facility (see Section 5.3), any calling DTE is allowed to request the fast select for a given virtual call by inserting such a request in the facility field of the call request packet followed directly by user data up to 128 octets. This "datagram" is delivered to the called DTE in the incoming call packet by the corresponding DCE as described in Table 4.26. Such a procedure looks as if it were of the connectionless-mode, however the call request/incoming call packets establish all necessary network resources via the PSPDN mechanism. Moreover, a called DTE is permitted to react to the incoming call by sending the call connected/clear request packet depending upon the demanded response parameter. This is of course at variance with the principles of the connectionless-mode, which is explained briefly in Table 4.19. The only explicit connectionless-mode network protocol implemented in public networks is the ISDN intranetwork protocol based on common channel signalling. Therefore our attention is drawn to ISDN network protocols.

As we have mentioned, the user-to-ISDN network layer protocol is performed on the D-channel with the signalling rates of 16 kbit/s or 64 kbit/s for the basic and primary access, respectively, in the packet switched mode. The protocol is based upon a set of *messages* rather than packets (in ISDN terminology) exchanged between the user (terminal equipment) and the ISDN at the S/T interface as usual. Messages, like packets, consist of a header and an information field (data, for example). They are classified in several types and their names differ slightly from those used in the X.25 network protocol. For example, instead of the call request packet the corresponding ISDN message is called a set-up message, clear request becomes a disconnect message, etc. For the full list of ISDN messages the reader is referred to CCITT recommendations (I.450/Q.930 and I.451/Q.931), where the complete access protocol at the network layer is also described. Here only a simplified view of the access to ISDN in terms of ISDN messages is given (Table 4.27). The table cannot give the content of messages (facilities, call progress signals) and does not cover all events which may occur during the transmission. On the other hand, it completes the comparisons between any switching mode and any access to public networks.

The ISDN is, however, much more sophisticated because it employs the results of development of new signalling systems: signalling systems No.6 (SS No.6) and No.7 (SS No.7). In particular, the latter, approved in the mid-1980s, forms the central nervous system of ISDN by its intranetwork protocols for signalling over the 64 kbit/s E-channel. It follows the OSI reference model in the two lowest layers (see previous chapters) and its network layer can be enhanced to the full OSI network capability.

Table 4.27.   An example of the user-to-ISDN protocol (S/T interface) at the network layer on the D-channel

Terminal equipment A (calling)		ISDN	Terminal equipment B (called)
Set-up	$\longrightarrow$		
	$\longleftarrow$ Call proceeding	Set-up	$\longrightarrow$
			$\longleftarrow$ Call proceeding
			$\longleftarrow$ Alerting
	$\longleftarrow$ Alerting		$\longleftarrow$ Connect
	$\longleftarrow$ Connect	Connect ACK	$\longleftrightarrow$
Information exchange $\longleftrightarrow$			$\longleftrightarrow$ Information exchange
Disconnect	$\longrightarrow$		
	$\longleftarrow$ Release	Disconnect	$\longrightarrow$
Release complete	$\longrightarrow$		$\longleftarrow$ Release
		Release complete $\longrightarrow$	

The three lowest layers of SS No.7 are collectively called the message transfer part and provide services similar to, but not quite equivalent to, the X.25 layer services. If the signalling connection control part is added as a network sublayer (the whole is then called the network service part), the full OSI network layer capability is achieved. As the connection-mode and the connectionless-mode are defined within this sublayer the D-channel becomes a universal transmission and communication means for packet switched data in any mode.

More detailed descriptions of the ISDN access and intranetwork protocols is beyond the scope of this book. The reader can find them in specialized monographs and handbooks on ISDN and in CCITT recommendations.

4.8 Higher Layers [14,23,31,33,47,52,53,54,58,72]

Higher levels are not usually included in public networks, with the exception of networks hosting non-data services (for example, telematic services), and, eventually, providing non-OSI services (packet assembly and disassembly, for example). Nevertheless, these layers form a constituent part of user DTEs and users should know how to design and implement communication control above transmission support. Here is a brief summary of them.

The transport layer is described in ISO standards (8072, 8073, 8602) as well as in CCITT recommendations (X.214, X.224, T.70). They employed works within ECMA (transport protocol ECMA-72) and the National Bureau of Standards (USA) which specified two classes of transport protocols close to classes 2 and 4 below.

Let us remind ourselves that the transport layer is intended to provide reliable transparent data transfer between end systems (the transport protocol is in most cases the lowest end-to-end protocol) and, together with the session layer, to provide the highest quality transparent dialogue between communicating partners at reasonable costs. Hence the transport layer must perform functions necessary to meet quality requirements, regardless of the quality of network services.

Although there could be many types of network services labelled by different quality values, only three types of network connections are recognized to date:

- Type A with an acceptable residual error rate and an acceptable rate of signalled (but not recovered) errors (see Section 7.1)
- Type B with an acceptable error rate but an unacceptable rate of signalled errors
- Type C with an unacceptable error rate

Note that the classification is based upon fuzzy values (acceptable/ unacceptable) and is user-subjective. For different users and different applications the limiting values vary to a great extent. Network users have five classes of transport protocols at their disposal, providing transport connections. These are:

- Class 0 (simple) – the least complex because it contains no facilities for error control, flow control and multiplexing. It is obviously intended for type A network connections. This class was developed by the CCITT for the needs of teletex (CCITT T.70)

- Class 1 (basic error recovery) – also developed by the CCITT as the transport protocol for PSPDNs whose three lowest layers are supported by the X.25 protocols. This class is intended for type B network connections and is a superset of Class 0 even though it performs only minimum error control but resets synchronism. Moreover, it enables expedited data transfer and acknowledgement
- Class 2 (multiplexing) – for use when several transport connections are to be multiplexed onto a single type A network connection; it can also solve the problems of flow control
- Class 3 (error recovery) – a superset of Class 2, intended for type B connections
- Class 4 (error detection and recovery) – provides error control which is capable of solving the problems arising when a type C network connection will be employed

As the classes cited above are not explicitly hierarchically arranged, the set of functions which they perform may indicate their power. Table 4.28 shows some functions by which the individual classes of transport protocols mutually differ (the inclusion of a function into a corresponding class is marked "x"). Most functions are known from previous chapters and only a few of them require explanation.

Thus, the concatenation provides for several transport blocks to be conveyed in one network service data unit and the separation is its counterpart. Transport numbering is recommended to be modulo 128, as shown in the table, but the optional extended numbering 2^{31} is allowed in Classes 2, 3 and 4 as well. Flow control makes use of a credit method while error detection is performed by numbering, block check sequence (checksum) and time-outs. Errors such as loss, corruption, duplication, disordering and misdelivery of blocks which occur as a result of the network's low quality of service are recovered by retransmissions. Splitting/recombining allows a

Table 4.28. Some functions by which the transport protocol classes differ

Function	Transport protocol class				
	0	1	2	3	5
Concatenation/separation		x	x	x	x
Numbering modulo 128		x	x	x	x
Expedited data transfer		x	x	x	x
Multiplexing/demultiplexing			x	x	x
Flow control			x	x	x
Network connection failure recovery		x		x	x
Resynchronization when a reset after failure recovery takes place		x		x	x
Block check sequence (checksum)					x
Retransmission					x
Resequencing of blocks					x
Splitting/recombining					x

transport connection to make use of multiple network connections to increase reliability and throughput.

Since the size of data fields in the PSPDN network layer data units (packet and service data units) is restricted (for example, the packet user data field should contain 2^K octets where K may vary between 4 and 12) and transport service data units may exceed these values, the transport layer performs the segmenting of data service units in shorter user data fields (fragments) and the reassembly to the original data message. The block (transport protocol data unit) is also predetermined in both length and structure: the header consists of a fixed part containing frequently occurring parameters (commands, credit, send sequence number, designation of the last fragment, and so on), and a variable part carrying less frequently used parameters (such as options, additional information, values of quality of service, checksum). The header may be followed by user data (normal, expedited). The length of the whole block is again recommended to be 2^K but with K between 7 and 13, and the user data field must respect the header length.

The principles of data exchange over a transport connection by means of a chosen transport protocol are similar to those applicable in the case of link and network layers even though there are differences in the names of commands and responses, in the parameters used, and in methods and functions. In particular, the data transfer exchange phase differs in classes of transport protocols according to the functions performed (see Table 4.28).

Let us consider only transport protocols providing the connection-mode service and using network connections. In that case the establishment of a connection involves the following: a user (in the transport layer terminology, an initiator: for example a transport entity which has initiated the communication) must first assign the transport connection being established to one or more suitable (as for quality of service) network connections. Then the connection request block is sent to a responder (the opposite transport entity) which responds by the connection confirm block or by the disconnect request (if the transport connection is refused). In the case of class 0, simultaneous two-way flow of user data blocks is performed without any error control (with the exception of the control of protocol errors) and flow control. The transport connection is released by either transport entity which sends the disconnect request block and gets the reply of disconnect confirm.

Instead of examples of transport protocol samples, an illustration of how standard protocols allow the construction of a superstructure of public networks is presented. Fig. 4.36 shows the three lowest layer protocols which may be applied in CSPDN, PSPDN and ISDN over the D-channel, and also

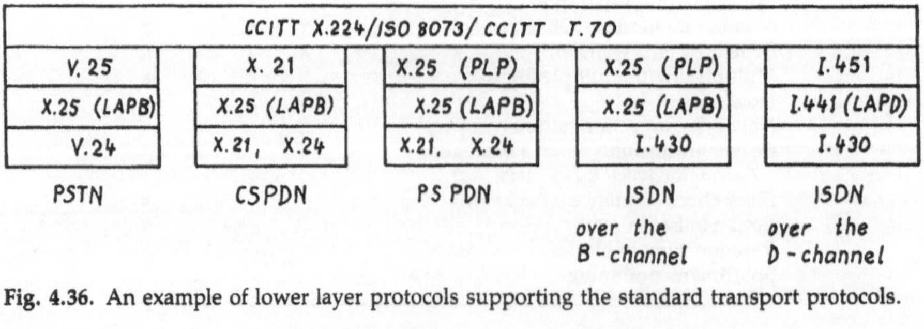

CCITT X.224/ISO 8073/ CCITT T.70				
V. 25	X. 21	X.25 (PLP)	X.25 (PLP)	I.451
X.25 (LAPB)	X.25 (LAPB)	X.25 (LAPB)	X.25 (LAPB)	I.441 (LAPD)
V.24	X.21, X.24	X.21, X.24	I.430	I.430
PSTN	CSPDN	PS PDN	ISDN over the B-channel	ISDN over the D-channel

Fig. 4.36. An example of lower layer protocols supporting the standard transport protocols.

for access to PDNs via the PSTN and ISDN over the B-channel. The examples are not exhaustive. For example, the X.25 link access procedure balanced (LAPB) can be replaced by the X.75 single link procedure (SLP), etc.

Beside transport protocols providing the connection-mode service supported by the network connection-mode service, other possible combinations are allowed and are already standardized. ISO 8073/Add 2 is designated for co-operation with the connectionless-mode network service while ISO 8602 solves the transport connectionless-mode with the support of an arbitrary network service (connection-mode as well as connectionless-mode). However, a deeper analysis of such cases is beyond the scope of this book and the reader is referred to the bibliography at the beginning of this chapter or to relevant CCITT and ISO documents (see Appendices 3 and 4).

The universal approach concerning mode is kept in higher layers, too. Session services and protocols have been elaborated by both ISO and CCITT (and ECMA). ISO dealt right at the beginning with wider user oriented aspects in ISO 8326 and 8327 and this concept has also been adopted by CCITT (X.215 and X.225). However, CCITT has concentrated more on session support for telematic services such as teletex, facsimile and videotex (for example, the session protocol T.62 is not fully compatible with the general session protocol).

The late 1980s illustrate the consensus between ISO and CCITT not only regarding the session layer services and protocols but also other layers of the OSI reference model. The close co-operation between the leading standardization bodies resulted in materially identical and technically aligned documents as seen in the X.200 series of CCITT recommendations, where differences between the corresponding documents are picked up and listed as appendices.

The presentation layer is often identified only by alphabets and codes. However, the presentation services and protocols have already been standardized and the remaining work consists mainly of harmonizing the message handling services of the X.400 series and other services having been in operation with the appropriate data presentation. Fortunately, the presentation and session services are almost independent and the session services are directly or in a modified form provided through the presentation layer. This fact simplifies the presentation service definition and presentation protocol specification.

CCITT covers within the application layer several telematic services such as message handling systems (X.400 series), document transfer and manipulation (T.400 series), teletex and Group 4 facsimile (T-series), and directory services (X.500 series). The application area is of course much broader and is the domain of ISO. It comprises user services such as job transfer and manipulation (ISO 8831, 8832), virtual terminal service (9040), general text communication (8505, 8883, 9065, 9066, 9072, 10021), file transfer and management (8571), computer graphics (7942, 8632, 8651, 8805, 9636) and others in the process of standardization. Their explanation is space consuming (the volume of valid documents already exceeds one thousand pages) and cannot be included here. The new documents that are continually appearing, not only internationally but also regionally and from various scientific and industrial organizations, should be the main sources for those who want to keep up with the state of the art.

5 ■ PDN Supported Services and Facilities
[8,14,22,72]

5.1 Basic Means and Basic Classification

In compliance with the general concept of service defined in Section 4.1, the services supported by PDNs imply primarily the fulfilment of users' requirements for data transmission, which is the transmission of signals between their DTEs. However, data transmission is only a means of satisfying the user's communication requirements, whether it is data communication or any other form of communication where data transmission is involved. A PDN is not the only telecommunication system capable of providing such services. On one hand DTEs can be linked by leased (permanent) circuits of the telegraph, telephone and broadband type or by switched circuits (physical or virtual) in the PSTN or private data networks, and, on the other hand, overall integration of services in the ISDN will comprise all standardized services as well as those services which make use of data transmission to fulfil users' requirements.

These considerations comply with the basic classification of telecommunication services into *bearer services* (or transmission services) limited to the transmission of signals between user-to-network interfaces and *teleservices* (complete telecommunication services) within which the network operator (telecommunication administration or RPOA) yields to the user's disposal the necessary terminal equipment as well.

From the economic point of view teleservices based on data transmission can bring to the operator of the PDN (or to those data transmission users who provide these services to other users) additional revenue (which is usually greater than the revenue from data transmission). Hence the terms "value added network" and "value added services".

A characteristic feature of subscriber services (telephone, Telex, telefax, teletex) is the existence on the subscriber's premises of equipment made available to him in order to access the service. This equipment is usually called the *subscriber station*. It allows him to communicate with other subscribers to the service on the general principle "every user with every other user". A necessary prerequisite (and task for the service provider) is securing compatibility of subscriber equipment as well as keeping the subscriber informed about the possibilities of the service and about other subscribers. This is done by issuing directories and by giving information about subscriber numbers.

The counterpart of subscriber services are hand-in services. The service provider receives the message from the user and takes charge of its telecommunication transmission and subsequent delivery to the addressee. An example of such a service is the telegram service (public telegraph service) and bureaufax. In contrast to subscribers, the users of hand-in services can be called *clients*.

To make the classification complete, it is necessary to apply the criterion of the direction of the main information flow (unidirectional and bidirectional, the latter called dialogue mode or – in the case of real time communication – interactive mode). If more than two users participate in telecommunication (point-to-point communication is extended to multipoint) unidirectional can be subdivided into centralized and decentralized multipoint. Centralized multipoint can be further divided into point-to-multipoint communication or broadcasting or – more generally – distribution (including selection mode) and multipoint-to-point communication or collection (including polling mode). A typical example of bidirectional multipoint communication is the conference service. Fig. 5.1 summarizes the criteria for the general classification of telecommunication services mentioned in this section.

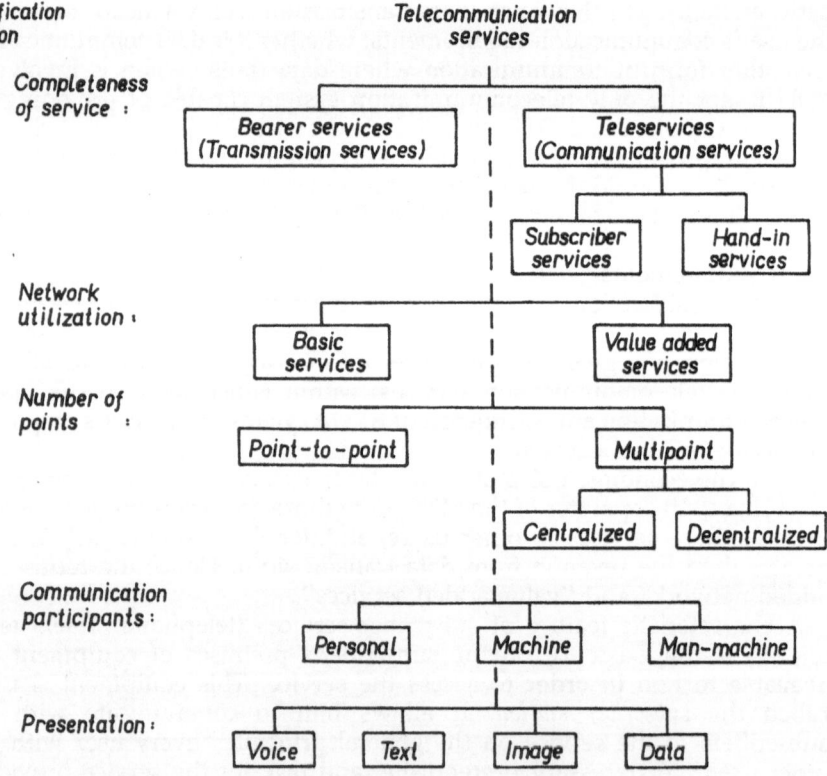

Fig. 5.1. General classification of telecommunication services.

The users of telecommunication services within a single network can be classified into user classes of service according to mutual accessibility and terminal equipment compatibility. Within a given service a user is offered various options called *user facilities*. Services and user facilities define the scope of user requirements which a service provider is capable of satisfying. The definition of a service is not complete without some objective determination of the quality of service, expressed by the degree of this satisfaction. In a broader sense, a service is codified by the price (codified by tariff) paid by the user.

5.2 Data Transmission Services

PDNs and ISDNs secure the transmission of signals between user DTEs by making complete end-to-end circuits (in the case of a CSPDN or its equivalent in the ISDN) available to its subscribers or by taking over data messages from a subscriber and forwarding them via the network to their addressees (in the case of a PSPDN or its equivalent in the ISDN). Hence the basic classification criterion is the principle of satisfying the users' data transmission requirements. The second criterion concerns the method of bit-serial transmission across the DTE–network interface, where distinction is made between start–stop and synchronous transmission. In start– stop transmission the sampling time base of the receiver is triggered by the reception of a start signal and stopped by the reception of a stop signal, and the samples correspond to the received bits, usually those of a character with a constant number of bits. In synchronous transmission the bit-sampling time base runs continuously and is synchronized by the received signal yielding an uninterrupted bit stream.

Another relevant classification criterion is the data signalling rate during the data transfer phase and call control phase. This criterion subdivides data transmission services into user classes of service standardized by CCITT recommendation X.1. The overall classification of data transmission services and their relation to user classes of service, listed in Table 2.1, is given in Table 5.1. If the data transmission service is provided by the ISDN, a special

Table 5.1. Correspondence between data transmission services (DTS) and user classes of service in PDNs and ISDNs (see Table 2.1)

Network	DTS	DTE operating mode	User class
PDN	Circuit switched	Start–stop	1, 2
		Synchronous	3–19
	Leased circuit	Start–stop	1, 2
		Synchronous	3–7, 19
	Packet switched	Synchronous	8–13
		Start–stop	20–23
ISDN[a]	Circuit switched	Synchronous	30 (19)
	Leased circuit	Synchronous	30 (19)
	Packet switched	Synchronous	30 (13)

[a] Valid for reference point S/T; for reference point R (with the use of terminal adaptors) all other PDN user classes can be implemented as well.

class of service (30) is used with a data signalling rate of 64 kbit/s for circuit switched and packet switched applications, making class 30 equivalent to classes 19 and 13, respectively.

In packet switching, DTEs with different data signalling rates can communicate with each other, the difference in speed being overcome by the storage of packet-constituting bits in the memories of packet switches or packet handlers in ISDN exchanges.

In the early days of PDNs the majority of DTEs operated in the start–stop mode over switched and leased connections. This mode was adapted to entering information by means of the keyboard and to the possibilities of mechanical terminals of the teleprinter type. At the top of this group is the class 1 DTE working with an 8-bit code plus one start bit and a stop element with two unit intervals corresponding to two bits, 11 bits per character in total. The data signalling rate of 300 bit/s (class 1) matches the maximum speed of a mechanical character printer (approximately 30 characters per second).

In addition, the start–stop group has to host further DTEs working with various data signalling rates (50 bit/s to 200 bit/s) and character formats 7.5 to 11a (where a is the unit interval). They form the second user class. Within this class, the following data signalling rates and code structures are recommended: 50 bit/s – 7.5a, 100 bit/s – 7.5a, 110 bit/s – 11a, 134.5 bit/s – 11a.

For user classes 3–6 operating in synchronous mode those data signalling rates which are common for data transmission over the PSTN, as recommended by the V-series recommendations, were used. The signalling rates 48 kbit/s and 64 kbit/s are compatible with CCITT recommendations for broadband modems (V.35, V.36) operating in the primary group band mainly for 12 telephone channels.

No code is prescribed for the data transfer phase of any class. In the call control phase of classes 1–7 and 19 the code of IA5 is obligatory, because the DTE and the exchange to which it is connected must speak a common language when setting up, supervising and clearing a connection. If the service provider wants to integrate the Telex service into his PDN, he has to apply a modified 50 bit/s option by using 50 bit/s and International Telegraph Alphabet No.2 (ITA 2) in the call control phase, too.

Interworking between a start–stop DTE and the PDN is defined by interface recommendation X.20. To make it possible for a user with a start–stop DTE laid out for data transmission over the PSTN to connect this DTE to the CSPDN, an alternative recommendation, X.20bis has been detailed for use on PDNs of DTEs which are designed for interfacing to asynchronous duplex V-series modems. A similar provision is valid for DTEs designed for interfacing to synchronous V-series modems. In the latter case recommendation X.21bis is used instead of X.21. The "bis" interfaces are also applicable when a DTE is accessing its PDN via the PSTN or via a leased telephone-type circuit.

Synchronous mode DTEs of the packet switched data transmission service in user classes 8–11 and 13 use the same data signalling rates as synchronous DTEs of the circuit switched data transmission service in user classes 4–7 and 19. Class 12 is designated for the access of synchronous DTEs to the PSPDN via the PSTN.

Start–stop mode DTEs of the packet switched data transmission service accessing the PSPDN via a packet assembly/disassembly facility (PAD) in compliance with recommendation X.28 work with data signalling rates and formats defined by user classes 20–23. User class 20 corresponds to classes 1 and 2. However, the formats are limited to 10 and 11 unit intervals. Class 21 has two data signalling rates. These are 1200 bit/s in the direction from the network to the DTE and 75 bit/s in the opposite direction – a combination suitable for communication between a keyboard/display terminal and a distant database, as in the videotex service.

In addition to the basic data transmission services mentioned above, PDNs provide other services. Examples of these are the provision of a multiplex DTE–DCE interface and of ports for atypical DTEs. An example of the former is time division of the 48 kbit/s bit stream of user class 7 into bit streams yielding user classes 3–6. Examples of the latter are DTEs which had been used in private data networks prior to the introduction of PDNs, ports for systems working with the binary synchronous communication (BSC, IBM) for example, and the synchronous data link control (SDLC).

The existence of private data networks prior to the appearance of PDNs imposes on PDN designers and operators the commitment to secure for the users of networks which have been incorporated into PDNs the same standard of services and operating conditions as before. A solution to this issue is the use of gateways or bridges between the private data networks and the PDN or the connection of the private network to the PDN via the standard subscriber interface. The latter solution applies mostly in the case of LANs, which constitute metropolitan area networks (MANs) when interconnected directly or by switching via a PDN in metropolitan areas or with unlimited range wide area networks (WANs). The principle of virtual calls and permanent virtual circuits in a PSPDN can be extended to set up virtual private networks (VPNs) within the PSPDN. The provision of these networks with autonomous management systems for the benefit of users can, according to the definition of service, also be regarded as a service.

5.3 User Facilities

The PDN or ISDN operator (telecommunication administration or operating agency) offers to the users of services in the user classes mentioned in Section 5.2 specific possibilities tailored to their needs. These possibilities are called *user facilities* and are made available to users under the assumption that their DTE complies with the requirements defined within this facility. This compliance must lend itself to identification by the network.

Since the leasing of circuits for data transmission is also regarded as a service provided by the network, the ways of utilizing leased circuits are user facilities. These ways correspond to the classification of leased circuits into point-to-point and multipoint. Multipoint leased circuits are further subdivided into centralized and decentralized.

Optional user facilities may be assigned for a contractual period or requested by the DTE on a per-call basis. User facilities can be grouped according to their reason for existence by means of the following criteria:

Table 5.2. Survey of relevant optional user facilities

Use facility	Circuit switched		Packet switched	
	Agreed period	Per-call basis	Agreed period	Per-call basis
Abbreviated address calling		x		x
Bilateral closed user group	x		x	
Bilateral closed user group selection		x		x
Call deflection selection				x
Called line identification		x		
Calling line identification	x			
Charging information		x	x	x
Closed user group	x		x	
Closed user group selection				x
Connect when free	x			
Date and time indication	x			
D-bit modification			x	
Direct call	x	x	x	
DTE inactive registration/cancellation	x			
Extended frame sequence numbering			x	
Extended packet sequence numbering			x	
Fast select				x
Fast select acceptance			x	
Flow control parameter negotiation				x
Hunt group	x		x	
Incoming calls barred	x		x	
Non-standard default packet sizes			x	
Non-standard default window sizes			x	
Network user identification (NUI) selection				x
Network user identification (NUI) subscription			x	
One-way logical channel incoming			x	
One-way logical channel outgoing			x	
On-line facility registration		x		
Outgoing calls barred	x		x	
Packet retransmission			x	
Reverse charging		x		x
Reverse charging acceptance	x		x	
Throughput class negotiation			x	x
Transit delay selection and indication				x

Table 5.3. Mutual accessibility of DTEs (see Fig. 5.2)

DTE	User facility	Connection possible	
		From DTE	To DTE
A	Closed user group I with outgoing access	B, D, E	B
B	Closed user group I with incoming access	A	A, C, D, E
	Closed user group II with outgoing calls barred	A	A, C, D, E
C	Closed user group II	B	D
D	Closed user group II with incoming calls barred	B, C	A, E
E	Outside closed user group	B, D	A

- Simplification and acceleration of user-to-DTE communication, enhancement of user comfort (direct call, abbreviated address calling, date and time indication, multi-address calling)
- Barring of mutual accessibility to preserve the advantages that users have in private networks (closed user groups, bilateral closed user group, barring of certain types of calls)
- Enhancement of network connectivity (redirection or deflection of calls, connect when free, waiting allowed, hunt group, DTE inactive registration, fast select)
- Identification facilities (network user identification, called line or calling line identification)
- Charging-related facilities (charging information, reverse charging, reverse charging acceptance, local charging prevention)
- Non-standard features or parameters (one-way logical channel, non-standard default packet size, packet retransmission)
- Facilities associated with flow control (flow control parameter negotiation, throughput class negotiation and assignment)
- On-line facility registration
- Facilities made available in the case of a DTE accessing the PSPDN over the PSTN or CSPDN (dial-in and dial-out access)
- End-to-end signalling (by D-bit modification)

Some user facilities apply only to CSPDNs, some only to PSPDNs, some to both these networks or to accessing PSPDNs via PSTNs or CSPDNs (see Table 5.2, which lists facilities in alphabetical order). The characteristics of the individual user facilities are described below, in the order of the categories mentioned above.

Direct call is the setting up of a switched connection (virtual or circuit switched) without an indication of the called subscriber's address, this address being stored in the memory of the network. It can be used even on a per-call basis by simply applying a time-out on the absence of normal selection.

The abbreviated address is to be loaded (together with the full address) into the memory of the exchange by the user in a predetermined way. Whenever the user applies the abbreviated address facility, he calls the full address by the abbreviated address (prefix). The user can delete the address in the memory when he no longer wants to use it.

Date and time indication is used in the CSPDN. It informs the users about the point of time when they have set up a mutual connection (year, month, day, hour and minute).

A multi-address call allows the user to transmit a message simultaneously to several addressees whose addresses had been made known to the exchange in advance. It is a facility used in the CSPDN.

Closed user groups (CUGs) are established to protect users by a system of access limitations from undesired communication with other PDN users, including those who are themselves members of some user group. Users not belonging to any group form the so-called open part of the network. Closed user group related user facilities are subject to the following classification (Fig.5.2 and Table 5.3):

1. Basic CUG (GUGa) with users belonging to one or more CUG

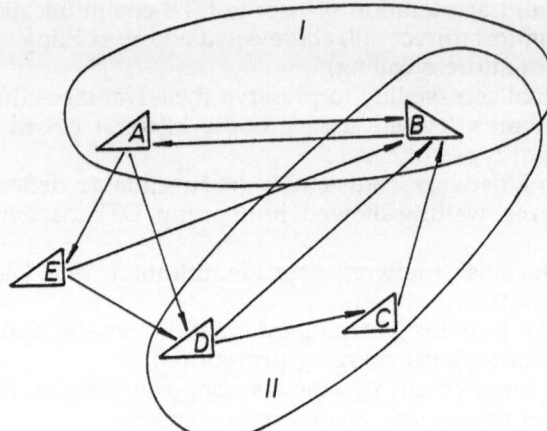

Fig. 5.2. Mutual accessibility of DTEs in closed user groups (CUGs) and DTEs not belonging to any CUG.

2. CUG with outgoing access (CUGb) is a kind of CUGa whose users have access only to the open part of the network and to CUG users with incoming access (CUGc)
3. CUG with incoming access (CUGc) is a kind of CUGa, where users accept calls from the open part of the network and from CUGb users
4. Incoming calls barred within a CUG is an additional user facility for individual users of groups CUGa, CUGb or CUGc
5. Outgoing calls barred within a CUG is an additional user facility for individual users of groups CUGa, CUGb or CUGc

If a user belonging to one CUG becomes a participant of another, the original CUG is his preferential CUG. If a user belonging to more than one CUG becomes a participant of a further CUG, one of his original CUGs will be regarded as preferential. Each user belonging to at least one CUG has either the user facility of CUGa or one of the facilities of CUGb or CUGc. If he is a participant of CUGb as well as CUGc it is up to him to decide whether he will grant priority to CUGb or CUGc. CUG user facilities are implemented by a system of inhibition code words and are based upon a checking operation during the call set-up phase.

A bilateral CUG permits a pair of DTEs who bilaterally agree to communicate with each other to do so, but – unless specially agreed otherwise – communication with all other DTEs is barred.

The call redirection facility allows the network, in certain circumstances, to redirect calls destined to the originally called DTE. The circumstances may be:

- The DTE is out of order
- The DTE is busy
- Systematic call redirection due to a prior request by the subscriber: with one alternative DTE or a list of alternative DTEs, with or without logical chaining. To preclude a call rambling in a closed loop the call set-up is limited in time. For example in a PSPDN a time-out of 200 s is applied

By means of another facility – the call redirection notification – the DTE to which the call is redirected is informed why the call has been redirected and is given the address of the originally called DTE.

The call deflection facility allows a DTE subscriber to deflect an individual, already established, call to another DTE. A notification similar to that about call redirection can be made.

Connect when free in a CSPDN or ISDN allows an incoming call to wait until the called DTE is capable of accepting it. This facility is contingent on the assent of the called subscriber given within the waiting allowed facility.

The hunt group facility, if subscribed to, distributes incoming calls having an address associated with the hunt group across a designated group of DTEs.

If a DTE inactive registration is made the calls to this DTE are barred. This contributes to the suppression of waste traffic in the network.

Fast select is a user facility in the PSPDN which allows short messages as well as the corresponding response messages to be transmitted via the network without virtual call set-up. This facility ostensibly replaces the connectionless-mode or datagram facility (one message per packet), which has been partly abandoned in modern PDNs (see Section 4.7.3).

The network user identification (NUI) related facilities enable a DTE in a PSPDN to provide information to the network for the purposes of billing, security and network management, or to invoke subscribed facilities. Calling line identification is a facility provided by the CSPDN which enables a called terminal to be notified by the network of the address from which the call has originated. A similar definition applies to called line identification.

Charging information is sent by the DCE (representing the network) to the DTE. The information makes it possible for the user to calculate the charge. Reverse charging is charging the called subscriber instead of the calling subscriber if the calling subscriber requests this facility and the called subscriber agrees by activating the facility reverse charging acceptance. The local charging prevention authorizes the network to prevent the establishment of calls which the subscriber must pay for. This is brought about by not transmitting incoming calls which request the reverse charging facility to the DTE and by ensuring that the charges are debited to another party whenever a call is requested by the DTE.

A number of user facilities is implied by the principle of packet switching. For example, the packet retransmission facility allows the DTE to request retransmission of one or several consecutive data packets from the associated DCE (representing the network). The one-way logical channels facility restricts the logical channel use to outgoing or incoming virtual calls only (even though the full-duplex transmission capability is retained in both cases).

Some user facilities give the user – who acquires them by request either for a contractual period of time or on a per-call basis – better conditions of service. For example, this is the case of non-standard default values. A default value is the basic value which the user is offered if he does not indicate a preference for a different value. The network designer can take into consideration the default values as those which are relevant for network dimensioning. Examples of a standard default value are packet size (maximum data field length) 128 octets, window size 2, and packet sequence numbering modulo 8.

Packet and window sizes affect flow, and the setting of their values can control the packet flow from DTEs into the PSPDN. Therefore, the sizes are called flow control parameters and their setting is done as a result of another user facility, the flow control negotiation between user and network operator (service provider).

A throughput class is a measure of a steady state throughput (in bits per second) that can be provided under optimal conditions on a virtual switched connection or on a permanent virtual circuit. The user has the possibility of throughput class negotiation and throughput user class assignment as user facilities. The throughput values are chosen from those which correspond to user classes of service. Throughput classes are considered independently for each direction of data transmission.

The on-line facility registration is a user facility which allows the user at any time to request registration of facilities or obtain current values of facilities by communication with the network.

The D-bit modification facility (see Section 4.7.3) applies to all virtual switched connections and permanent virtual circuits and provides for DTE-to-DTE communication (for end-to-end acknowledgement) by setting the binary value of the D-bit, not only in the data packet.

Automatic test and network information stations for all user classes and facilities can be made available as a special user facility.

5.4 Support of Complete Telecommunication Services

These services, which in CCITT terminology are also called *teleservices*, as opposed to the *bearer services* dealt with in Section 5.2, provide a complete capability, including terminal equipment functions, for communication between users according to established protocols. The implementation of teleservices in a PDN upgrades this network to a *value added network* and the teleservices themselves assume the character of *value added services*. Added value is an appropriate attribute because teleservices usually bring in more revenue to the service provider. Typical teleservices procured by a PDN or its equivalent in the ISDN are: electronic message handling (also called electronic messaging) according to recommendation series X.400 and F.400, directory services according to series X.500 and F.500, electronic data interchange (EDI), electronic fund transfer, interconnection of LANs and provision of VPNs, and most of the CCITT defined telematic services (series F.350), including videotex (series F.300), telefax 4 (F.184), teletex (series F.200) and teleconference (series F.700).

Interworking with Telex (for example, via the Telex/teletex conversion facility which can reside in the PDN hosting the teletex service) can also be regarded as a teleservice. Another teleservice is the support of telecommunication between personal computers including distant program loading and intermediating PC access to the videotex service.

Almost every domain of human activity can be supported by a data communication system based upon a PDN. The function of this system is usually implemented by a data processing and data logging computer interworking with remote general-purpose or customized terminals (banking terminals, point-of-sale terminals, teller terminals, educational terminals, logistic terminals, for example) and workstations (sophisticated terminals where sent and received data are a by-product of the workstation's activity). Beside these human-operated equipments the PDN can support various fully automatic data acquisition systems and effectors in industrial processes and monitoring of the environment.

5.5 Support of PDN-Type Services and Facilities by the ISDN

In principle, the ISDN should be capable of providing the same services and user facilities as dedicated telecommunication networks, including PDNs. The advantage of the ISDN over the conventional telephone network is in its creation of a fully digital environment extending into the subscriber's premises.

The system of ISDN services is described by CCITT series I.200 recommendations. The services provided by the network respond to customer needs and generate requirements on network capabilities which in turn must be matched to terminal equipment capabilities. The throttling effect of economic aspects on one side and the enhancing effect of the development of technology on the other have to be considered. Services in the ISDN environment are subdivided into bearer services, teleservices and sup-plementary services. Additional services correspond to user facilities in PDNs. The counterpart of user classes of services in PDNs are service categories in the ISDN.

Bearer services assume access in reference points S and/or T of the subscriber interface (see Chapter 2). They are of either the circuit-mode or the packet-mode type.

The first stage of the circuit-mode bearer service standardization concerns the narrow-band ISDN with a 64 kbit/s basic bit rate and its multiples: 384, 1536 and 1920 kbit/s unrestricted, 8 kHz structured. Multiple subrate information streams are to be multiplexed into 64 kbit/s by the user. A transparent access to a PSPDN or any other X.25 network via the above mentioned reference points is also envisaged. User information is transferred over the B-channel, signalling is provided over the D-channel.

The packet-mode bearer service categories involve virtual call and permanent virtual circuit, connectionless-mode and user signalling. The framework for additional packet mode bearer services has been elaborated.

The first teleservices to be supported by an ISDN were telephony, teletex, telefax 4, mixed mode, videotex and Telex.

In broadband ISDN the following service classes are envisaged: con-versational services, messaging services, retrieval services and distribution services with or without presentation control by the user.

A supplementary service is defined as one which cannot be provided as a stand-alone service. Similar to user facilities, supplementary services can be assembled into groups:

- Number identification
- Call offering
- Call completion
- Multiparty
- Community of interest
- Charging
- Additional information transfer

The association of supplementary services to basic PDN-type teleservices supported by the ISDN is given by Recommendation CCITT I.250.

6 ■ DTE Access to Public Services and Networks

6.1 Significance of Access Specifications
[22,33,40,44,53]

The concept of DTE access involves all means and methods of communication between a DTE and the network by which the considered PDN service is provided. The data transmission service can be either of the switched or the leased circuit type. The network, if public, can be a CSPDN, PSPDN or ISDN. Contrary to the common use of the term, access here covers both directions of communication set-up, outgoing as well as incoming (the same applies of course to the directions of transmission). Access can be achieved:

- By direct connection of the DTE to the PDN or ISDN
- By switched connection of the DTE to a PDN or ISDN via an intermediate network of another type (including PDN, PSTN, Telex and ISDN with or without a terminal adaptor, see Chapter 2)

CCITT recommendation X.10 specifies categories of access for DTEs to data transmission services provided by PDNs and ISDNs. These categories are listed in Tables 6.1 and 6.2. A special kind of access is the access of non-packet-mode DTEs to a PSPDN via a packet assembly and disassembly (PAD) facility. A survey of examples of DTE access to PDNs and ISDNs is given in Fig. 6.1. The DTE communicates with the service-providing network in full-duplex or half-duplex mode. Both channels of the connecting circuit can have the same (symmetric circuit) or different (asymmetric circuit) data signalling rates. An example of an asymmetric circuit is user class 20 with bit rates 1200 bit/s from DCE (network) to DTE and 75 bit/s in the opposite direction. Incoming and outgoing access to and from a data service providing a network is based upon the utilization of terminal equipment, transmission and switching facilities of telecommunication networks which had been in existence prior to the introduction of the PDN or ISDN. The connecting of non-compatible DTEs to the PDN creates the necessity of inserting convenient conversion facilities between these DTEs and the PDN or ISDN proper. This applies to the majority of access to PSPDNs (where the conversion facility is the PAD) and to the ISDN where the conversion facility is the terminal adaptor (TA).

Table 6.1.   Categories of direct access to PDNs

Type of PDN	Type of DTE	Category[a]	Data signalling rate (bit/s)	Interface DTE–DCE
CSPDN	Start–stop	A1	50 to 200	X.20, X.20bis
		A2	300	X.20, X.20bis
	Synchronous	B1 (S1)	600	
		B2 (S2)	2 400	X.21, X.21bis
		B3 (S3)	4 800	and X.30
		B4 (S4)	9 600	
		B5 (S5)	48 000	
		B6 (S6)	64 000	
PSPDN	Start–stop	C1	110	
		C2	200	
		C3	300	X.28
		C4	1 200	
		C5	75/1 200	
		C6	2 400	
	Synchronous	D1	2 400	
		D2	4 800	X.25, X.31
		D3	9 600	
		D4	48 000	
		D5	64 000	
		(T1, U1)	2 400	
		(T2, U2)	4 800	
		(T3, U3)	9 600	(X.25, X.31)
		(T4)	48 000	
		(T5)	64 000	

[a] Categories and recommendations in parentheses refer to direct access established through ISDN facilities. T categories are related to the B-channel, U categories to the D-channel.

Transmission facilities for access are based upon symmetric pairs in the local telephone networks and upon analogue and digital transmission systems in different media (symmetric, coaxial and optical fibre cables, microwave links, connections via telecommunication satellites). Access via telecommunication networks other than that which provides the data transmission service is performed with the aid of interworking units or – more generally – interworking functions. The interworking is standardized by CCITT series X.300 recommendations (see Chapter 4). Standardization of access is closely related to standardization of DTE–network interfaces (series X.20 and X.30 for PDNs and I.310 to I.470 for ISDNs).

6.2 Access Through PADs [5,22,44,51]

If the PSPDN is to be accessed by non-packet oriented DTEs (at present mostly cheap start–stop terminals of the teleprinter type) the PAD function must be provided by the network. To date, only PADs for connecting

Table 6.2. Categories of switched access to PDNs

Type of PDN	Access network	Type of DTE	Category	Data signalling rate (bit/s)	Interface DTE–DCE
CSPDN	PSTN	Start–stop	G1	300	
		Synchronous	I1	600	
			I2	2 400	
			I3	4 800	
			I4	9 600	
	ISDN channel B		J1	600	
			J2	2 400	X.21, X.21bis,
			J3	4 800	X.30
PSPDN	CSPDN	Start–stop	K1	300	
	PSTN		L1	110	
			L2	200	X.28
			L3	300	
			L4	1 200	
			L5	75/1 200	
			L6		
	CSPDN	Synchronous	O1	2 400	
			O2	4 800	
			O3	9 600	
			O4	48 000	
			O5	64 000	
					X.32
	PSTN	Synchronous	P1	1 200	
			P2	2 400	
			P3	4 800	
			P4	9 600	
	ISDN channel B	Synchronous	Q1	2 400	
			Q2	4 800	
			Q3	9 600	X.25, X.26
			Q4	48 000	
			Q5	64 000	

start–stop DTEs are internationally standardized (CCITT recommendation X.3).

The main problems of start–stop DTE interworking with the PSPDN lie primarily in the incompatibility of the format of transmitted data. Start–stop DTEs are capable of working only with characters equipped with start and stop elements and possibly also with one or more parity bits for error control, whereas a PSPDN recognizes only packets of a prescribed format. Therefore it is necessary to insert a conversion facility, which is capable of assembling the received start- stop character signals into a packet of assessed format, between the PSPDN and the start–stop DTE and in the opposite direction to disassemble the packet received from the distant DTE into a string of start–stop characters.

This implies also the control of character flow from the start–stop DTE because a data packet has a limited data field and the start–stop DTE cannot be constrained as to the length of transmitted messages. A convenient

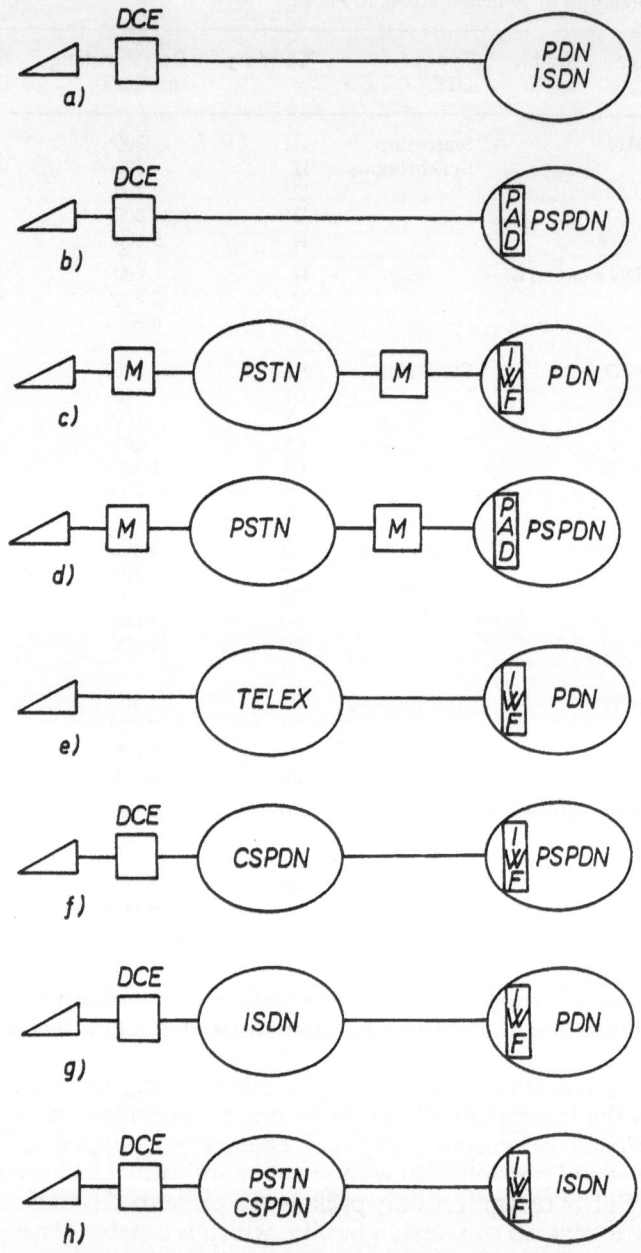

Fig. 6.1. DTE access to PDNs and ISDNs. IWF, interworking function; M, modem. **a** Direct access of packet oriented DTEs to a PDN or ISDN; **b** direct access of nonpacket oriented DTEs to a PSPDN via a PAD; **c** switched (dial-in and dial-out) access of a packet oriented DTE through the PSTN to a PSPDN; **d** switched (dial-in and dial-out) access of a nonpacket oriented DTE through the PSTN to a PSPDN; **e** switched access through the Telex network to a PDN; **f** switched access through a CSPDN to a PSPDN; **g** switched access through an ISDN to a PDN; **h** switched access through a PSTN or a CSPDN to a PDN.

separation sign in the sequence filling in the data field of the packet is the carriage return character (CR) for teleprinter type terminals.

For data terminals employing IA5, the separating function is assigned to the following control characters: end of transmission (EOT), enquiry (ENQ), acknowledge (ACK), delete (DEL), horizontal tabulation (HT), form feed (FF), or any alphanumeric character of this alphabet. It is also necessary to take into consideration that some start–stop DTEs operate in the character-echo mode – the emitted characters do not appear directly on the display of the terminal but only after circulation through the loop DTE– network–DTE (these two DTEs being identical).

In the opposite direction (from PAD to DTE) it is not possible to send an unlimited sequence of characters because it would not match the format which the DTE supports (for example, into one line of text) – hence the need to insert appropriate formatting characters (carriage return, line feed) into certain positions of the data sequence and, in the case of mechanical teleprinter-type terminals, filling characters for the terminal to manage formatting operations. In addition, it is necessary to add the start and stop element and possibly also parity elements. DTEs access a PAD by direct or switched connections; a PAD is a part of the PSPDN (Fig. 6.2), usually integrated into the PSPDN switching node.

Because of the great variety of DTEs, it is not economical to design a special PAD for every DTE type. The PAD must be sufficiently universal to be capable of interworking as far as possible with all manufactured and operating start–stop terminals, therefore it must have incorporated parameters – the so-called *PAD parameters* – whose values are set and changed by telecontrol from the packet-oriented DTE or from the opposite PAD. As consecutive setting of the values of all parameters would be time-consuming with current DTEs, a family of DTE parameter values are set by a single command. If many DTEs are connected to a PAD, the PAD performs the functions of a concentrator.

PAD functions can be summarized as follows:

- Packet assembly from DTE character sequences
- Extraction of user data from received packets (that is, disassembly of the user data field of packets)
- Handling virtual call set-up and clearing
- Resetting and interrupt procedures

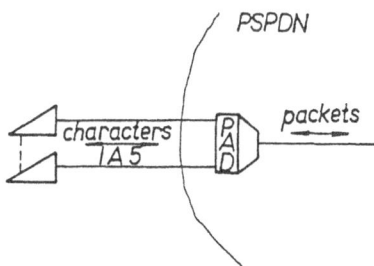

Fig. 6.2. Interworking between a start–stop DTE and a PSPDN through a PAD.

- Generation of service signals
- Forwarding packets into the PSPDN as soon as they have been formed (as soon as they are full) or an idler timer expires
- Transmission of start–stop data characters including start elements, stop elements and possibly also parity-check elements
- Editing PAD command signals
- Setting and reading valid PAD parameter values

These are the so-called basic PAD functions. In addition, PAD performs other optional or user selectable functions, such as selection of standard profile (setting of parameters and their values) and automatic detection of data signalling rate, code, parity and operational characteristics).

Fig. 6.3 shows the principle of PAD functioning. For each connection with a DTE there must be a buffer store in the PAD as well as a table of parameter values of the connected DTE. Since the PAD is a part of the PSPDN and co-operates closely with its subscriber, its characteristics must be very well standardized. There are three CCITT recommendations (called triple X) meeting this requirement: X.3 defines PAD characteristics, X.28 the protocol on the interface DTE–PAD, and finally X.29 the protocol between PAD and a distant packet-mode DTE or another PAD. PAD parameters are on the one hand related to PAD functions and, on the other hand, they express the variety of non-packet-mode DTEs connectable to the PSPDN. Each parameter has a defined set of values which it can assume and, because the PAD must be capable of reading these values, a convention has been made on their coding in the decimal system (with binary transmission). PAD parameters are divided into those which affect the behaviour of the PAD and those which affect the behaviour of the connected DTE. These are some examples of parameters (with the associated decimal coding) effective in the PAD:

- Existence (1) or non-existence (0) of echo
- Time-out between two consecutive characters from the DTE after the expiry of which the PAD completes the forming of the packet and sends it into the network (0 to 255 indicate the time-out value in milliseconds)
- Choice of procedure after reception of the interrupt signal, i.e. signal of

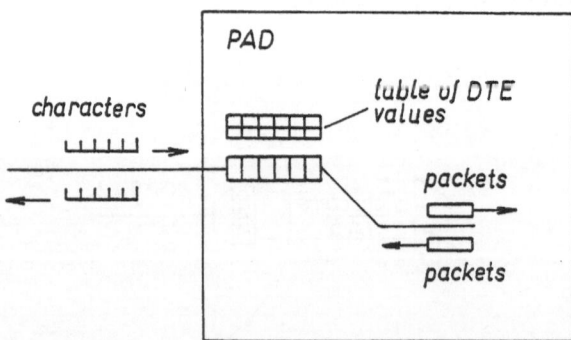

Fig. 6.3. Functioning principle of a PAD: conversion of character strings into packets and vice versa, according to stored PAD parameter values.

condition 0 for more than 135 ms which is generated by depressing the "break" key (0 – no activity, 1 – transmission of the interrupt packet, 2 – 1reset, 4 – indication of break PAD message, 8 – escape from data transfer state, 16 – discard output data for the DTE); padding (insertion of filling characters after the character of carriage return or new line for the DTE); number of inserted padding characters (1 to 255)

- Data signalling rate of DTE: 0 – 110 bit/s, 2 – 300 bit/s (mandatory selectable values, other values optional); parity treatment: 0 – no parity checking or generation, 1 – parity checking, 2 – parity generation

Examples of parameters effective in the start–stop DTE are:

- PAD recall (0 – not possible, 1 – character DLE, 32 to 126 – using any character defined by the user)
- Control of PAD service signals and PAD command signals (0 – no signals, 1 – signals in standard format, etc.)
- Data flow control of the PAD by the terminal by sending the device control character DC1 (DTE ready to receive data) and device control character DC3 (not ready): 1 – existence of this function, 2 – non-existence
- Interpretation of editing PAD service signals (EPADSS) (0 – no EPADSS, 1 – EPADSS for printing terminals, 2 – EPADSS for display terminals, 3 – EPADSS using one character from the range of IA5, 8 and 32–126)
- Ancillary device control: 0 – no use of this control, 1 – use of DC1 (ON) and DC3 (OFF)

On the DTE side a PAD handles protocols of the physical layer and link layer. According to the means of transmission the following ISO standards and CCITT recommendations apply:

- In the case of an analogue direct connection: between DTE and modem with interface circuits according to recommendations V.24 and V.28 and with the connector ISO 2110, modem to network interface: V.21, V.22 and V.23 (current start–stop terminals work with a data signalling rate not exceeding 1200 bit/s)
- In the case of an analogue switched connection: in addition to the above mentioned interface V.24 an additional interface according to V.25 or V.25bis for manual or automatic selection
- In the case of a digital connection (direct or switched via a CSPDN) the DTE–DCE interface complies with X.24 with electrical characteristics according to X.26, X.27 or V.28 with connector ISO 4903, or in the case of access to the CSPDN via an analogue circuit with X.20bis modems

In the link layer IA5 characters are used and the transmission obeys X.28. On the PSPDN side the PAD works according to X.25 in the three lowest layers.

A specific feature of the incorporation of PADs into the PSPDN is the need for partial control on the session and presentation layers. The very principle of PADs demands an entirely different way of session set-up and clearing towards the considered DTE accessing the PAD than towards the packet mode DTE or another PAD. Therefore X.29 applies, which on one hand modifies the network layer control as compared with X.25 and on the other hand specifies the control of session and presentation.

For this purpose the so-called PAD messages (4 bits) are specified. They are responses to commands from the distant packet-type DTE and allow for setting or reading parameter values. The structure of a packet with a PAD message is illustrated in Chapter 4. The Q bit (eighth bit) in the first octet indicates by Q = 1 that this packet contains a PAD message. The D bit (seventh bit in the same octet) is set to 0. The message itself is limited to 128 octets.

The procedures between the PAD and the distant packet mode DTE or another PAD are not seen by the considered start–stop DTE.

6.3 Access Via TAs [8,40,42,44,69]

DTEs of PDN user classes of service (see Section 5.2) access the ISDN at reference point R via the TA which gives the DTE the same conditions as when it is connected to a PDN. On the other hand the ISDN sees these DTEs as if they were ISDN-type terminals which connect directly to network termination at reference point S/T. An exception to this is class 30 (user class specific to ISDN) where the DTE accesses the ISDN as other ISDN terminals at reference point S/T.

Situations of insertion of TAs are according to recommendation X.30 for non-packet-type DTEs with interfaces X.21, X.21bis, and X.20bis, and according to X.30 for packet-type DTEs with the X.25 interface.

The functions of TAs for non-packet-type DTEs are rate adaption, protocol conversion, call offering procedure on a multi-terminal configuration, ready for data alignment, retransmission of call progress signals, handling of exceptional situations and flow control. TA functions for packet-oriented DTEs are rate adaption and multiplexing (Section 4.5.3), signalling conversion, synchronization and maintenance. They are provided for the access through both the B-channel (64 kbit/s) and the D-channel (16 kbit/s).

TAs for connecting DTEs with interfaces to CCITT V-series modems at the S/T reference point are dealt with in recommendation V.110.

6.4 Direct Access [7,22,70,72,74]

This type of access is performed by a permanent connection between the DTE and a point of access to the network which provides the data transmission service. On the basis of a user facility called *multilink procedure* a single DTE can communicate with the network over several data links simultaneously.

The choice of the connecting circuit depends upon the user class of service according to CCITT recommendation X.1 which determines the signalling rate and transmission mode. The applied technique of transmission depends upon the distance from the DTE to the point of access to the network. The distinction between local and distant is roughly the same as in Telex, that is, it is given by the distance coverable within a local telephone area of the PSTN. If the DTE is within the same local telephone area (area served by a local telephone exchange) as the corresponding PDN or ISDN exchange, the usual methods of digital data transmission over one pair of conductors in a local cable apply: low level and low impedance data transmission sets, baseband modems and in the case of asynchronous DTEs of low bit rates (up to 300 bit/s) methods of teleprinter signal transmission via local networks. For high bit rates, 64 kbit/s and 2.048 Mbit/s, methods of PCM signal transmission over local networks are used.

If the DTE and the network access point are not within the same local telephone area, greater distances are covered by the use of modems at both ends of the connecting circuit (CCITT V-series recommendations) or by channels in FDM systems (as in voice frequency telegraphy according to CCITT R.30-series) or in TDM according to R.100-series and X.50-series (see Section 4.5.3). The use of the statistical multiplexer and associated demultiplexer is important for the utilization of transmission capacities in direct access (see Table 6.1). This allocates tributary channels only to active inputs so that the sum of nominal input data signalling rates is greater than the data signalling rate on the multiplex bearer channel.

6.5 Switched Access [22,34,41,44]

In the case of switched access a switched connection is established between a DTE and the accessed network (PDN, ISDN) in a network of another type (for example, the PSTN, the Telex network, the CSPDN, any private data network) and the ISDN. The reason for accessing the data service-providing network through a switched connection is that the establishment of a direct connection would be uneconomical (possibly because of the excessive distances) or that the switched connection serves as a standby route for the case of direct connection interruption or failure. The latter case frequently occurs, since, from the point of view of system redundancy, the access circuit is the weakest and least reliable link in the communication chain.

To give the DTEs connected to the access network the possibility of setting up a switched access connection, the accessed network must have special access points (access ports) addressable by the accessing network. The functions of interworking between the accessing and accessed network are concentrated into network interworking units.

To make the switched access connection equivalent to direct access connection, both directions of access have to be considered: incoming access (dial-in) and outgoing access (dial-out). The services provided for users with DTEs connected to access networks are called DTE services: service for unidentified DTEs, service for identified DTEs and service for customized (personalized) DTEs. A customized DTE is billable, has an X.121 address registered with the PDN, and is provided with a service which is in many aspects tailored to its requirements.

In connections involving a switched access circuit it is desirable to ensure end-to-end identification; a mere identification of the access point is not sufficient. An exception to this requirement could be a calling DTE in the case of reverse charging. End-to-end identification is required for the calling party to ensure that it has reached the called party, for initiating the necessary charging procedure and for categorization compliance testing. Categories of switched access according to CCITT recommendation X.10 are given in Table 6.2. The interface between a packet-oriented DTE accessing its PSPDN via a PSTN, an ISDN or CSPDN is standardized by X.32.

Each international connection can be regarded as having switched access because – from the point of view of the called DTE – it has been set up through at least one foreign network playing the role of an access network. Where direct internetwork connections do not exist, it is necessary to set up calls via third networks.

6.6 Numbering Plans and Directory Services
[32,38,46,48]

The necessity of allocating unambiguous addresses to members of any group who have to communicate with each other is generally known (see Section 4.2.4). This holds especially for telecommunication network users. For these users the network provides a domain of network service access point (NSAP) addresses (CCITT X.213, annex A) controlled by an addressing authority. An NSAP address comprises a string of up to 40 decimal digits. It consists of an initial domain part (IDP) followed by a domain specific part (DSP). The IDP consists of a two-digit authority and format identifier (AFI) followed by the initial domain identifier (IDI). The NSAP can thus be represented as follows:

NSAP address = IDP + DSP = (AFI + IDI) + DSP

For many reasons, existing telecommunication networks use decadic numbers (in principle, binary octets or even letter characters can be used as well) as addresses for their users in an agreed numbering plan. With the growth of international traffic, especially since its automation (such as the elimination of manual switching), national telecommunication numbering plans had to be fitted into a global international numbering plan.

Whereas a numbering plan defines numbering principles, the subscriber numbers themselves and their correspondence with subscriber names accompanied by postal addresses and possibly also by relevant attributes (in the so-called "Yellow Pages") are contained in directories. Directories are made available to subscribers in printed form (books), by means of directory enquiry services (nowadays mostly computer aided) run by operators or without operators (electronic directory according to CCITT series X.500), and last, but not least, by enabling the network user to communicate with distant directory data bases (a videotex application).

The basic idea of a numbering system within a telecommunication network involves allocating numbers to subscribers connected to a certain exchange. The numbering (in advanced systems) may not necessarily be identical with the numbering of lines connected to that exchange. The higher level of number allocation is related to the numbering hierarchy of exchanges and even networks, national as well as international.

The creation of a comprehensive and sophisticated international numbering plan for PDNs according to CCITT X.121 fits into the NSAP address system described above: with AFI = 36 (two digits) and IDI up to 14 digits, the IDP has up to 16 digits. Since the maximum length of the total NSAP is 40 decimal digits, the length available for the domain specific part (DSP) amounts to 24 decimal digits. Similarly, the ISDN numbering plan (according to E.164) with AFI $= 44$ (two digits) and a 15 digit IDI has IDP length 17 digits, and hence the maximum DSP length is 23 decimal digits. The following principles have been agreed as a basis for the international numbering for PDNs:

- The international PDN subscriber number is to determine only the specific DTE–DCE interface and to identify a country and a network, if several PDNs exist in the same country
- Where more than one PDN exist in a country, it should not be mandatory to integrate the numbering plans of the various networks
- The number of digits comprising the code used to identify a country and a specific PDN in that country is the same for all countries
- A DTE–DCE number is unique within a country (national data number); it forms part of the international data number which is unique on a worldwide basis (see Fig. 6.4)
- The international PDN numbering plan makes provision for the interworking with DTEs on the PDNs with those on PSTNs, in Telex networks and in the ISDNs

A data network identification code (DNIC) occupying the first four digits of the IDI for PDNs is assigned to:

- Each PDN in a country
- Non-zoned services, such as the public mobile satellite system
- A PSTN or ISDN with connected DTEs
- A group of PDNs within a country
- A group of private data networks connected to PDNs within a country

A DNIC has four digits, the first three representing the data country code (DCC), and the fourth being the network digit (see Appendix 2). Countries with more than 10 data networks have more country codes (the USA has seven, for example).

As to the first digit, the world is divided into seven zones:

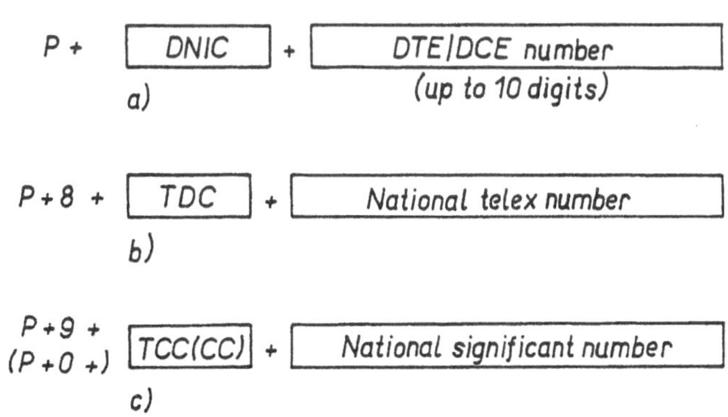

Fig. 6.4. International X.121 format. DNIC, data network identification code; P, prefix; TDC, Telex destination code; TCC, telephone country code; CC, country code, as defined in E.163. **a** International data number; **b** international Telex number; **c** international telephony/ISDN number, P + 9 + TCC – analogue interworking, P + 0 + CC – digital interworking.

1. World's oceans – system INMARSAT
2. Europe and Greenland
3. North America
4. Asia
5. Australia, New Zealand and the Pacific islands
6. Africa
7. South America

Digits 8, 9 and 0 are used as escape codes to access the Telex network, PSTN and ISDN, respectively (the latter two are regulated in the short term by X.122), and are not part of a DNIC. Their use is evident from Fig. 6.4, defining the international X.121 format.

A similar philosophy applies for access from other networks to the PDN. It is based upon recommendations E.163 with the numbering plan for the international telephone service, and E.164 with the principles of a universal numbering plan for the ISDN era (see Appendix 2). Though a unified global E.164 numbering plan for all telecommunication networks would be the ideal long-term solution, the existence of other than ISDN networks (CSPDNs, PSPDNs, Telex) makes this idea impractical. The long-term solution of this problem lies in the scenarios of numbering plan interworking laid down in E.166. Fig. 6.5 presents a selection of these long-term scenarios concerning PDNs and ISDN-provided data transmission services. The numbers of CCITT recommendations above the DTE symbol refer to the DTE–network interface of that DTE, those above the arrow to the called address (B) and those below the arrow to the calling address (A).

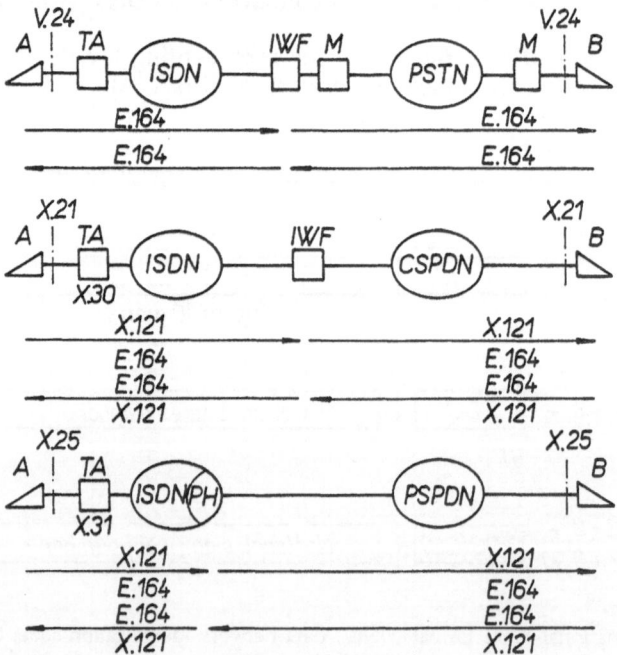

Fig. 6.5. Numbering plan interworking for DTEs connected to different types of networks.

7 ■ Network Performance

7.1 Two Approaches to Network Performance Assessment [25,66]

The performance of any system should always be examined from two aspects. A typical user is not concerned with how the provider achieves the meeting of demands upon a particular system service or how the internal structure of the system is designed. This also holds for communication systems and networks. The network user is focused on services consisting of the performance of particular applications (electronic fund transfer, file transfer, text communication, for example), on perceivable effects, objective measurements and intelligibility. From the service provider's point of view network performance should enhance network development, planning, operation and maintenance.

In order to distinguish those aspects, two categories of performance measures have been introduced. The network proper is evaluated by *network performance* (often abbreviated to NP) whereas users are rather more interested in the *quality of service* (QOS) provided by the network. Therefore network performance expresses the capability of a network to provide services (or more exactly, to perform functions supporting these services) and quality of service determines the degree of the service user's satisfaction.

Although either valuation measure relates to a single collective effect, for such complex systems as public networks a single figure of merit cannot give a comprehensive picture. Therefore, a set or vector of measures has to be found.

There are many measures which could be used to evaluate network performance and quality of service. Nevertheless, an appropriate selection is necessary in order to be independent of various applications, networks and services. With proper choice they can be specified irrespective of network internal design and protocols used and should be applicable to circuit switched networks and packet switched networks, to connection-mode as well as connectionless-mode. In addition, they have to be measurable and verifiable at well-defined boundaries or points.

The convention regarding these boundaries and points is the most crucial demand. This is because of the existence of different boundaries recognized in real networks and their models. A very natural approach is to designate the DTE–DCE interface (or more exactly, its line of demarcation) as the performance point. However, this point is, according to Section 4.4, a point of delivery of the data transmission service only and thus only the quality of data transmission can be defined there. Moreover, the DTE–DCE interface is protocol-dependent. As performance refers to services in general, the points of delivery of the particular services have to be considered. Fortunately, public networks consist of, in most cases, only the three lowest levels and hence the network layer service is identical with the service of a network. The reference points between which the quality of service of a network is to be determined are the network service access points. If a network provides network connections, the value of the quality of service of the network applies to a complete connection, when measured at either end of the connection. It is the same at both ends, and this is in fact true even in cases of interworking.

Fig. 7.1 shows the points where the network quality of service is observed. There are, however, two possibilities. The first possibility is to measure the quality of network service of the whole interconnection including end systems. We can call this the quality of OSI network service. The second possibility is to search the reference points within the network among the layer entities through which the network can be accessed. In the case of PDNs this point could be inside the network layer entities of DCEs and correspond to the quality of data transmission service. In general, the values of quality of the OSI network service are different from those of the quality of the data transmission service because of the operation external to the network. This operation, performed by the network service provider, may have the effect of either deteriorating or improving the quality of the data transmission service. This effect, and hence also the relationship between network internal and network external quality of service components, is part of the responsibility of the network service provider outside the network.

Quality of data transmission service is in fact equivalent to network performance. It is observed at the network boundary (at the DTE–DCE interface) and is based on events and states at connection element boundaries, protocol specific signals or protocol data units, for example. The quality of the OSI network service is measured and determined by making reference to primitive events at service access points which are independent of network processes and events supporting these processes.

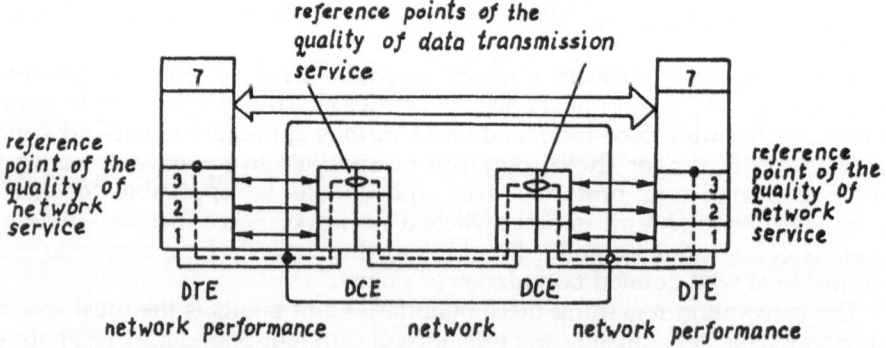

Fig. 7.1. The relationship between quality of network service and network performance.

Let us return to performance measures. The ISO and CCITT documents employ the term *performance parameter* though it can cause confusion with protocol parameters and parameters accompanying network services and user facilities (see Chapters 4 and 5). Nevertheless, in order to avoid introducing new terms differing from common and widely used ones we adhere to the term performance parameters and not performance measures. The same holds for the names of performance parameters which are the subject of the next section. Almost all performance parameters are of a random nature, most are often unknown, so their interpretation must take this fact into account. Statistics provides us with the mean value, the variance or its positive square root – standard deviation, and eventually, various percentiles. The recommended statistical characteristics for network performance are the mean value and the 95th percentile (a value of network performance not exceeding a probability of 0.95).

For quality of service parameters five values are recognized:

- A target value desired by the calling user
- The lowest quality value agreeable to the calling user
- An available value which the network provider is willing to provide
- A selected value to which the called user agrees
- A default value mutually understood and conveyed between network user and network provider

The first two values determine an applicable range of performance values.

The user oriented quality of service parameters provide a valuable framework for network design but they are not necessarily suitable for network operation and maintenance. Similarly, the network performance parameters more or less determine the quality of service observed by users without necessarily being described in a way that is meaningful to them. Therefore, both sets of parameters are needed and the values must be quantitatively related if a network is to be effective in serving its users. However, their definition should make mapping of values clear in cases where there is not a simple one-to-one correspondence between them.

Two approaches to evaluation led the CCITT to define the "common language" which would mutually relate with them. CCITT X.140 introduced the so-called *general parameters* based upon protocol independent events. Beside their definitions, the cross references with circuit switched and packet switched network service parameters and the quality of the OSI network service are given. We shall follow the general parameters in the next subsection.

Table 7.1, similarly to Table 4.1, gives references to CCITT and ISO documents concerning network performance and quality of the OSI network service.

Table 7.1. Survey of international standards concerning performance

General considerations	CCITT X.140, I.350
Circuit switched network performance	CCITT X.130, X.131, I.352
Packet switched network performance	CCITT X.134, X.135, X.136, X.137
Quality of the OSI network service	CCITT X.213, ISO 8348

7.2 General Parameters [22,25,58,66]

The search for appropriate performance parameters began in the 1970s, and some proposals have since become standards of national validity (American Standard X3.44/1974, British Standard BS 5208/1976). These parameters were intended for two-station communication systems only and did not fit in with multistation systems and networks. A newer and deeper analysis, taking into consideration complex computer communication systems and networks, was accomplished in May 1978 by N.B. Seitz and others (NTIA report, 78–4). Some results were demonstrated and verified by the performance evaluation of the ARPA network, thus forming the basis for US Federal Standard 1033, which was approved a year later. At the same time, these results were transferred to the CCITT where they were registered among the questions entrusted to study groups for the study period 1980–1984. The work, however, was not terminated in this period.

As indicated in the previous section we shall follow the general parameters recommended in CCITT X.140. References to other terms are included in order to avoid confusion and to guide the reader through other materials which are not fully compatible.

For comparability and completion a two-dimensional classification has been proposed. First, the communication process is divided into activities, termed functions. In order to avoid confusion with the functions introduced in Chapter 4 we shall call them *phases* because of their protocol independence.

Three phases are recognized. Access or selection begins upon the issue of an access request and ends when data transfer is enabled to start or is just starting. An example of access is the call set-up in the connection-mode.

User information transfer begins on completion of access, and ends when the end of transfer or the beginning of disengagement takes place. It includes all transmission, storage, switching, processing and media conversion operations performed on user information (user data).

Disengagement is the third phase associated with each participant in a communication process. It begins with the issue of a disengagement request and ends when the network resources dedicated to this communication have been released (call clearing, for example).

The second division concerns performance criteria and, again, three criteria are recognized. These are speed, accuracy and dependability (or, alternatively, inserveability or refusal). These express, respectively, the delay rate, degree of correctness of user data and degree of certainty the phase may proceed with.

In general, such a classification gives rise to nine entries, eight of which are illustrated in Table 7.2. Notice that two entries of the 3×3 performance matrix contain several parameters because of the complexity of performance evaluation. As the rows of the 3×3 matrix represent distinct phases and the columns represent exclusive outcomes, we can assume that each entry is independent. Moreover, as we shall see later, all selected parameters are mutually independent in the sense that none can be derived from the others, and therefore they are called *primary performance parameters*. They describe performance during periods when service is available, that is, in the absence of service outage. Hence, availability parameters are expressed by the

Table 7.2. 3 × 3 performance matrix

Phase	Performance criterion		
	Speed	Accuracy	Dependability
Access	Access delay	Incorrect access probability	Access denial probability
User information transfer	User information transfer delay	User information error probability	User information loss probability
	User information transfer rate	Extra user information delivery probability User information misdelivery probability	
Disengagement	Disengagement delay	Disengagement denial probability	

frequency and duration of periods of unavailable service, when service outages occur.

An associated two-state model forms the basis for the overall service availability evaluation (Fig. 7.2). A specified availability function compares the values of certain primary parameters with corresponding outage thresholds to classify whether the service is available (no service outage) or unavailable (service outage) during scheduled service time. Thus, availability performance parameters are derived from primary parameters and may be termed *secondary*.

The remainder of this section gives a brief summary of general performance parameters, primary as well as secondary. More detailed definitions with respect to network services (circuit switched, packet switched), service modes (connection-oriented, connectionless) and networks (PDN, ISDN) can be found in linking documents to X.140 (see Table 7.1).

The criterion of speed is described in four parameters. These are *access delay, disengagement delay, user information transfer delay* and *user information transfer rate*.

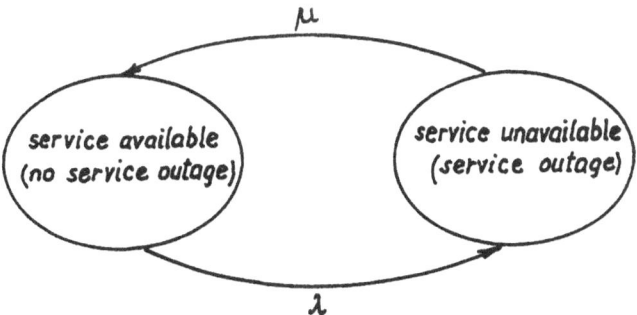

Fig. 7.2. Basic availability model.

Access delay is the value of elapsed time between an access demand and successful access. The successful access outcome is indicated either by the network issuing the demand accepted message ("ready for data" in the CSPDN, "call connected" in the PSPDN, "connect" in the ISDN), or by the fact that at least one bit of user data is put into the network (in networks providing the connectionless-mode, for example). The access delay, like many other performance parameters, is divided into user-dependent (caused by the DTE, for example, between the incoming call packet received and call accepted packet sent in a PSPDN), and network-dependent components. The latter can also be divided, if need be, into smaller well defined subcomponents.

Disengagement delay is the value of elapsed time between the start of a disengagement attempt made by a particular user and the successful disengagement of that user. The outcome of successful disengagement is indicated either by the network issuing clear confirmation, or by the fact that the user is able to initiate a new access.

User information transfer delay is the value of elapsed time between the start of transfer and successful transfer of a specified user data unit. The size of the user data unit must always be specified, this being the size of the data message in a CSPDN, or of the user data packet in the PSPDN. Moreover, this delay refers only to error-free data units. Transfer is initiated when the data unit is physically present in the sender facility of the network and the network has been authorized to transmit it. Similarly, the end of transfer is defined by the fact that the data has arrived at the destination user facility with the notification that the data unit is available for use. The definition seems to be complicated, however, as it precludes events like data unit misdelivery or transmission without permission.

The *user information transfer rate*, or, more generally, the *throughput*, is the number of successfully transferred data units per unit of time. The definition requires some comment. This parameter is most often expressed in bits per second, although other units (octet/s, character/s) are permitted as well. User data is hidden in protocol data units so that the number of user bits is the number of bits contained in data fields. Bit stuffing, error control, commands, identifiers, etc. are always excluded. User information transfer rate is a term of recommendation X.140 but others widely used are throughput (X.213, ISO 8348), information rate (Q.931) and information transfer speed (I.350). From their definitions a slight difference in meaning is evident and therefore they should be applied individually.

Throughput values are affected by such factors as data signalling rate, protocol data unit size, window size, throughput class, time-of-day and day-of-week. To eliminate these factors the throughput capacity is defined as maximizing the throughput over all combinations of user facility parameter settings under a statistically constant load. This means that each report has to specify the conditions under which the value has been obtained or is prescribed.

Instead of listing all factors and giving more precise definitions, an example of a throughput capacity report, taken from X.135 might be more useful:

For this connection the network throughput capacity is at least 4.1 kbit/s. The capacity was measured using two 9.6 kbit/s access circuit sections, data link layer window sizes of 7, packet layer window sizes of 2, and 128 octet user data fields. No additional virtual connections were present on either access circuit section. The capacity was measured during the busiest hour of the weekday. The average data packet transfer delay during the measurement period was 500 ms. The precision of the throughput measurement is plus or minus 0.1 kbit/s.

The criteria of accuracy and dependability are joint with probabilities of events occurring during communication expressed statistically as ratios of outcomes which actually estimate the true probability values (under assumptions well known from the theory of probability). Thus, the incorrect access probability is the ratio of total access attempts that result in incorrect access ("wrong number") to total access attempts.

Access denial (also termed network blocking) probability is the ratio of the total number of access attempts with access denial to total access attempts. User blockings where an access attempt fails as a result of wrong action of the user are excluded.

Disengagement denial probability is the ratio of the total number of disengagement attempts that result in disengagement denial to the total number of disengagement attempts.

The remaining four accuracy parameters form components of the *residual error rate* (RER). This common parameter is defined as the ratio of all incorrect, lost, extra (duplicate) or misdelivered user data units to all data units transmitted or received. It is defined for any kind of unit, which could be a bit, octet, character, frame, packet or fragment. However, this information must accompany the value (the bit RER). Residual refers to errors caused by influences during transmission, storing and processing, and by control faults and persisting after an error control function involved in the protocol in question has been performed.

The relationship between all events which may occur during communication and cause errors is illustrated in Fig. 7.3 by means of a Venn diagram. The total number of transmitted and received data units are designated N_T and N_R, respectively, and N_X out of N_R data units are extra or duplicate. N_T is partitioned into N_L lost data units, N_M misdelivered data units, N_E incorrect data units and N_S error-free data units.

A transferred data unit is defined as incorrect when one or more bits are in error, i.e. inverted, or when some, but not all, bits are lost or extra (not present in the original data unit). Hence, the user information error probability is $N_E/(N_E + N_S)$.

One limiting case of user information error probability is the bit error rate when the unit consists of a single bit. On the other hand, if the bit error rate concerns a sufficiently long user bit string an approach based upon the time portion related to erroneous bits seems to be expedient. By the choice of the 1 s time interval as a sample the number of all bits in the sample is just equal

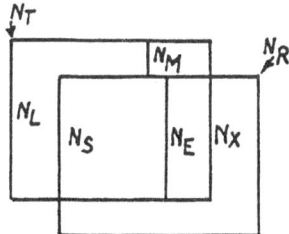

Fig. 7.3. The relationship between events determining the components of the RER.

to the value of data signalling rate so that the percentage of error seconds or error-free seconds can be determined. The latter measure is recommended and abbreviated as %EFS. A similar measure based upon 100 ms intervals gives the percentage of error-free deciseconds (%EFDS).

The definition of the other accuracy and dependability parameters related to the data transfer phase is obvious. The extra user information delivery probability is expressed by the ratio of the total number of extra data units to the total number of data units received by a destination (N_X/N_R). The user information misdelivery probability is given by the ratio of the total number of misdelivered data units (delivered correctly or incorrectly to a different destination) to the total number of data units transferred between a specified sender and destination ($N_M/(N_E + N_S + N_M)$). Finally, the user information loss probability can be given approximately by the ratio of the total number of data units lost due to the network to the total number of data units transmitted (N_L/N_T). Data units undelivered as a result of user refusal, by the flow control mechanism for example, are not taken as lost data units. Then the overall residual error rate is:

$$RER = (N_E + N_L + N_X + N_M)/N$$

where

$$N = N_T + N_X = N_R + N_M + N_L$$

The secondary parameters are based upon the notion of service outage. It is clear that a service outage includes any period during which the user is unable to elicit any response from the network (the network is completely deaf). However, the network can provide a service unacceptable to the user because of, for example, a poor value of residual error rate or throughput.

There are several approaches to determine the threshold values of primary performance parameters which would decide whether the obtained parameter value is acceptable or not. One follows the target values and allows their relative worsening: for example, the throughput threshold may be set between one-third and one-tenth of the target value, or the RER threshold is the square root of the target value, etc. Another approach, applied according to CCITT X.137, recommends certain availability decision parameters for particular types of PDNs, and associates their values with the absolute values of thresholds, where, for example, the threshold of the bit RER is authoritatively set to 10^{-3}. If the value of the decision parameter is equal to or better than the defined outage threshold, the performance relative to that parameter is considered acceptable. On the other hand, if the value is worse than the threshold, the performance relative to that parameter is considered unacceptable. The availability or unavailability of the service can be determined just from relative values of decision parameters. For example, if all decision parameters are acceptable, the service could be regarded as available. If, however, one or more decision parameters are unacceptable, the service would be unavailable.

In that manner two states can be defined (see Fig. 7.2) and it remains to measure the mean durations of individual states and the mean number of transitions between the states. In Fig. 7.2 two values are assigned to transitions: λ, representing failure rate, and μ, representing restoral rate. The well-known and often-applied availability parameters are the mean time between service outages (MTBSO) and the mean time to service restoral (MTTSR). The former is the average duration of any continuous available service time interval and obviously equals $1/\lambda$; the latter is the average duration of the unavailable service time interval (= $1/\mu$). The two parameters can be united to a common parameter – service availability A as the ratio of the aggregate time during which satisfactory service is, or could be, provided, to the total observation period (A = MTBSO/(MTBSO + MTTSR)), or its complement – service unavailability U (U = MTTSR/(MTTSR + MTBSO)). These parameters are often expressed as percentages.

CCITT X.140 involves, among availability, the user information transfer denial probability which evaluates the availability during the user information transfer phase with regard to the deteriorated values of throughput and residual error rate only.

The general parameters just surveyed seem for the time being to be appropriate for network performance evaluation. Since the general parameters are not fully identical with particular network performance parameters, there exist some differences in their names and definitions. Table 7.3 presents a certain relationship between general parameters and network performance parameters, evaluating various networks and their services. It compares general parameters whose values have been specified for CSPDNs, PSPDNs and ISDNs providing data transmission services on the circuit switched basis (CS-ISDN) and packet switched basis (PS-ISDN). The OSI network service parameters are also included. The relationship does not identify the parameters: some of them are interdependent but not identical. For more precise definitions the reader is encouraged to read the relevant documents.

A user, however, may ask for an additional evaluation, hardly expressible by numbers. Some of them are emerging: security (service capability to prevent unauthorized masquerading, monitoring, manipulation of user data), costs, priority, etc. Costs and data protection are considered the most important and are beginning to be more or less worked out. We shall deal with them in the following sections.

7.3 Tariff Policies [22,29,39,55,62]

The tariffs of telecommunication services follow the general tariff principles practised to a great extent by most PTT administrations and defined in CCITT recommendations of the D-series and in CEPT resolutions approved within the European Community. These principles may be summarized briefly as follows:

- Tariffs should reflect a reasonable return on the capital invested (profit)
- Service users should as far as possible obtain equivalent services for the same charges (harmonization)
- The charging policy should be comprehensible to the service user (understanding)
- Charge levels should lead to optimum demand for and optimum utilization of services
- Tariff policy should be as far as possible adapted to the social structure and general development of society

Table 7.3. Relationships between general parameters and various network performance parameters

General parameters	CSPDN	PSPDN	CS-ISDN	PS-ISDN	OSI network service
Access delay	Network post selection delay	Call set-up delay	Connection set-up delay; alerting delay	Virtual circuit set-up delay	Network connection establishment delay
Incorrect access probability		Call set-up error probability	Connection set-up error probability	Virtual circuit set-up error probability	Network connection establishment failure probability
Access denial probability	Blocking probability	Call set-up failure probability	Connection set-up denial probability	Virtual circuit set-up denial probability	
User information transfer delay		Data packet transfer delay	Propagation delay	Data packet transfer delay	Transit delay
User information transfer rate		Throughput capacity		Throughput capacity	Throughput
Residual error rate[a]		Residual error rate; reset stimulus probability; reset probability; premature disconnect stimulus probability; premature disconnect probability	Degraded minutes; severely errored seconds; errored seconds	Residual error rate; reset probability; reset stimulus probability	Residual error rate; network connection resilience

Disengagement delay	Network clear indication delay	Clear indication delay	Disconnect delay; release delay	Virtual circuit clearing delay	Network connection release delay
Disengagement denial probability		Call clear failure probability	Premature disconnect probability; connection clearing denial probability	Virtual circuit clearing denial probability; virtual circuit premature disconnect probability; virtual circuit premature disconnect stimulus probability	Network connection release failure probability
Service availability		Service availability	Network availability	Network availability	
User information transfer denial probability					Transfer failure probability
Service outage duration		Mean time between service outages	Network capability outage duration	Network capability outage duration	

[a] For the sake of brevity the general parameters of user information transfer accuracy, including the user information loss probability, are summed up under the residual error rate.

These (and other) ambitious and partly conflicting objectives are respected by almost all network providers when setting up tariffs for telecommunication services.

First, let us examine network cost. It may be divided into capital expenditures (cost of hardware, installation, maintenance), expenditures for the necessary software and personnel expenditures. Hardware involves not only equipment for switching nodes or exchanges but also network equipment, such as multiplexers, concentrators, PADs, DCEs or NTs on the users' premises, and of course transmission means (wires, metal cables, fibre optics). The cost of transmission media has been and remains very high, despite optimum utilization of trunks due to multiplexing and packet switching. However, subscriber lines always form a major item. Expenditures for transmission media amount to more than one-third of the remaining costs.

Software expenditures include the cost of basic software supplied by the manufacturer as well as the cost of application software supplied by the network provider. Software includes the necessary tools to support operation, maintenance and future planning. Software tools are mostly provided by network providers but are generally charged for.

Personnel expenditures are not negligible: they are often comparable with the transmission media costs. They cover monthly salaries of the planning, operation, maintenance, management/administration and installation staff. Sometimes they involve the costs of training the different groups.

Installation itself requires capital, too: either in the operational environment (fail-safe power supplies, wiring and grounding, air conditioning, rooms for technical equipment and for operational staff) or new buildings fully equipped with the necessary accessories.

The following factors are relevant to charging for data communication services, including those provided by PDNs and ISDNs:

- Type of service and user facilities
- Data volume and/or call duration
- Distance
- Traffic intensity (heavy/weak) or time of utilization

While there is enough experience with charging data transmission services provided by CSPDNs and PSPDNs, which is available both from the providers and the users, at the time of writing there were only scarce results in this domain as far as ISDNs are concerned (only from ISDN pilot operations). The ISDN era will not differentiate between voice and non-voice (data) usage because of the digitization of all information transmitted, so traffic charges should not be discriminatory. However, as we shall see below, the PDN tariff structure differs somewhat from that of the PSTN. Harmonization of the two approaches during the parallel development of ISDN and PDNs is no simple task. In order to avoid speculation we shall deal here only with tariff structures of PDN services for which publicized tariff rates have been exploited.

Similar to other telecommunication services, the charges for PDN services comprise:

- Access charges for the connection to the PDN that are paid for every connection independent of the traffic exchanged
- Traffic charges for the utilization of the PDN

Access charges consist of an initial fee (non-recurring charge) for the installation of a subscriber line and its termination or DCE, and a subscription rental payable at certain intervals of time (for example, monthly) until the subscription is cancelled. The initial fees may or may not depend upon data signalling rate, but, as a rule they are the same for broader ranges of rates (up to 300 bit/s, 1200 kbit/s to 9600 bit/s, over 9600 bit/s) while the subscription rentals do depend both on the data signalling rate and on user facilities. In all PDN services providing temporary connections, both CSPDNs and PSPDNs providing virtual circuits, calls are billed for each call attempt on the DTE side, even unsuccessful attempts. The unsuccessful attempts, of course, do not include events caused by the network provider (network overload, network failure). However, the billing of unsuccessful attempts discourages automatic repetition (automatic polling) and avoids network congestion.

The *utilization* of an established data connection in PDNs is charged according to the call duration as well as to the data volume transmitted. In PSPDNs the call duration may be measured from the moment of terminating the transmission of the "call connected" packet until the packet "clear request"has been received by the switching centre which calculates charges and quotes traffic values. The purpose of charging the call duration is to curb a user in uncalled-for engaging of connections and thus to prevent idle overloading of the network.

The charge for call duration, which is usually measured in minutes, may depend in CSPDNs upon the data signalling rate, the time of day within which the connection is used (periods of heavy traffic, usually in working hours, and periods of weak traffic which can be further divided into periods of out of working hours and periods of night hours and holidays) and upon the distance between the two DTEs. The latter can again be expressed discretely by several (2–5) distance zones similar to those assessed in the PSTN. Charges in PSPDNs are usually distance-independent, or a distinction is made only between local and trunk services. Higher charges are levied for international calls. On the other hand, the data volume is taken into account (in particular in data and interrupt packets) and is measured in segments (S). One segment is recommended to contain 64 octets (512 bits). The charge per unit of volume may depend additionally upon the time period, similar to the duration charges in CSPDNs, and upon the total data volume. For example, the first 100 kS (100 000 S) are charged at quadruple the basic rate, the second 100 Ks at double the basic rate and each additional 100 Ks is made at the basic rate. In addition, charges can be collected for the use of various items (such as PADs), for services (permanent virtual circuit (PVC), switched virtual circuit (SVC)) and user facilities. The surcharges are billed recurrently (usually monthly) regardless of the degree of utilization. The charge for a permanent virtual circuit is an example of this.

Table 7.4 shows typical components of charges for services provided by PDNs and factors which influence their amounts. The table does not include additional charges which bear upon the access into the PDN via other switched networks (PSTN, Telex, another PDN, ISDN) and charges for international data connections. All charges and surcharges can be found in PTT regulations and rate tables.

The assessment of charges can considerably influence the behaviour of users, revealed, for example, in a preference for certain user facilities or

Table 7.4. Components of charges for services provided by PDNs

Charge	Component	Factor influencing the amount of charges		
		CSPDN	PSPDN	
			PVC	SVC
Initial	Subscriber line	User class	User class	User class
Subscription rental	Subscriber line	User class	User class	User class
	PAD	–	Number	–
	Circuit	–	Number	–
	User facility	Assigned	Assigned	Assigned
Traffic	Call	Number Distance Duration Day of week Time of day Data signalling rate	–	Number Duration Time of day
	Data volume	–	–	Overall volume Time of day
	PAD	–	–	User classes 1–7
	User facility	Requested	–	Requested

services. On the other hand, the user is given the possibility of estimating the amount of money which he will have to pay for various applications of teleprocessing supported by PDNs. Let us illustrate this by an example.

Figs 7.4 and 7.5 depict the charge components together with the amounts of charges expressed in hypothetical units. These amounts are derived from actual tariffs valid for existing PDNs and should do justice to their mutual relations. They cannot, of course, be found in any official rate table.

Let us consider, as an example, a bibliographic service supported by start–stop terminals with a data signalling rate of 300 bit/s in local operation during working hours (period of heavy traffic), and let us compare its costs in a CSPDN and in a PSPDN. Suppose that one log-on to a bibliographical database queries 20 extracts. Gaining five abstracts requires 17 minutes and 260 segments have to be exchanged. Applying the amounts of charges taken from Figs 7.4 and 7.5 we come to the conclusion that the cost for PSPDN services (even if PADs are involved) amounts to less than one-third (20.2) of those using CSPDN (65.1).

For the message handling service by means of 2400 bit/s terminals (synchronous or packet for CSPDN or PSPDN, respectively) the profit is still higher. For example, one log-on to a mailbox, the review of headers of all messages, the complete reading of four messages of lengths not exceeding 500 characters, and the transmission towards the mailbox of one message, also not in excess of 500 characters, result in an application taking 5.5 minutes and in which 72 segments are exchanged. Using the CSPDN, the user has to pay 25.30 (see Fig. 7.4), compared with only 2.71 for the PSPDN (see Fig. 7.5), which is only one-tenth of the cost. Neither figure takes into account costs needed for call establishment because, according to our rate tables, the charges for both PDNs are the same (0.50).

We can, however, show an opposite example. When communicating by means of word processors at 2400 bit/s higher demands are laid on data volume than on call duration and the CSPDN proves advantageous. The transmission of one page of 2000 characters or 4 S taking 20 s over the CSPDN costs only 0.77 (during the period of weak traffic this could be as low as 0.37) but via the PSPDN costs 1.10 (0.38).

The comparison of the cost of data communications can be illustrated graphically. Taking an eight hour working day and the period of heavy traffic as a basis, the data signalling rate 2400 bit/s, and using Figs 7.4 and 7.5, we

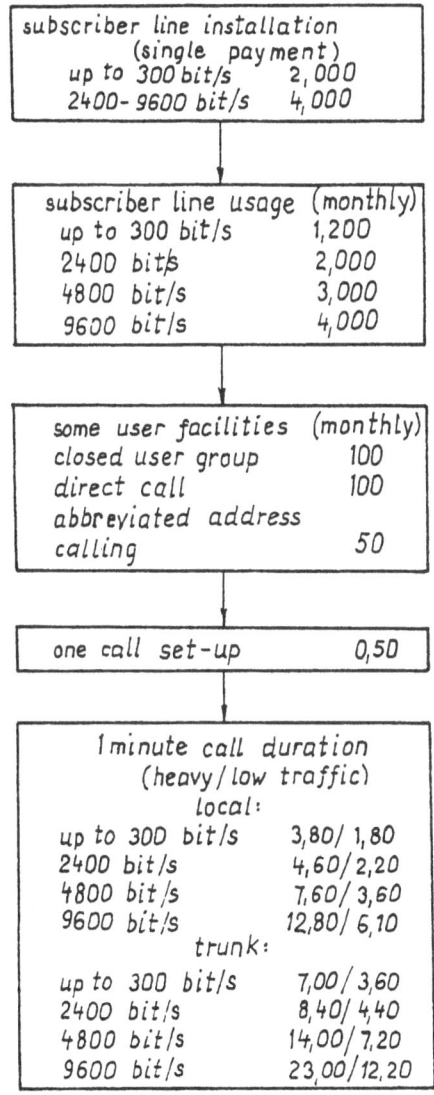

Fig. 7.4. Example of charges for services provided by a CSPDN.

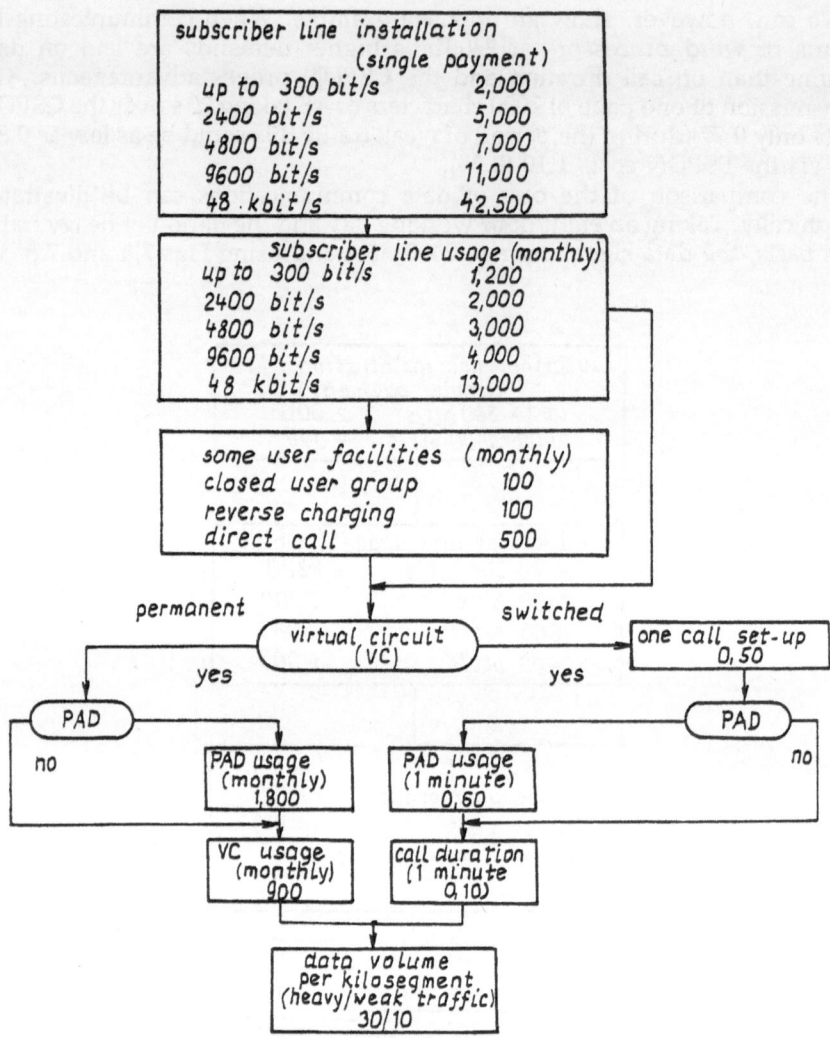

Fig. 7.5. Example of charges for services provided by a PSPDN.

obtain Fig. 7.6, which shows the dependence between daily charges and data volumes for services provided by leased lines, a CSPDN and a PSPDN. Note that leased lines (including PVCs in PSPDNs) are to be preferred if a certain limit of data volume is exceeded, whereas for small data volumes (regardless of the number of calls) the CSPDN within short distances, and the PSPDN irrespective of distance are advantageous.

Each user is given the possibility of applying the rate tables published by the corresponding PTT administration to determine the optimal areas of usage of data transmission services. Optimization is a fundamental prerequisite for increasing the efficiency of user data communication systems within the PDN as well as ISDN environments.

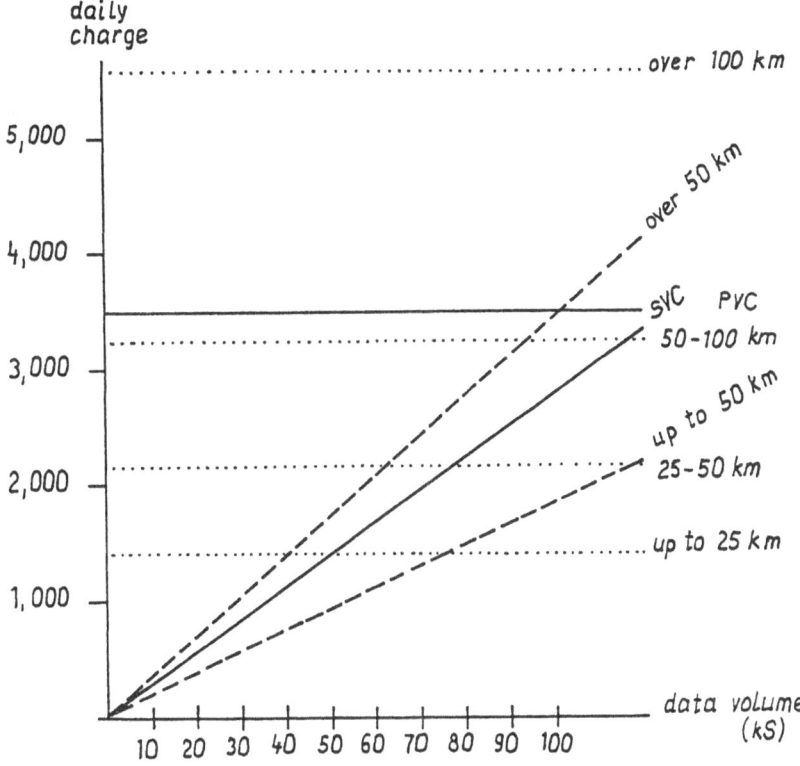

Fig. 7.6. Comparison of charges for data communication over a leased lines (dotted lines); b CSPDN (dashed lines); and c PSPDN (solid lines).

7.4 Network Security [14,16,20,28,41,45,53,60,64,72]

Security in data processing and communications has been an issue for several decades, but practically applicable results have been published only recently. The problem of security has become critical in computer networks and remote data processing since only mainframes with a limited selection of local users are capable of being physically protected by walls, locked doors and guards.

Security in general is something like the property of being free from evil. In the OSI environment the term *security* is defined more precisely: a means of minimizing the vulnerabilities of assets and resources where an asset is understood to be anything of value and vulnerability to be any weakness that could be exploited to violate a system and/or the information it contains. A network together with DTEs (hosts, terminals, etc.) are, by their nature, shared systems. DTEs are shared by users; switching centres (network nodes) and lines (circuits) are all shared by many different information flows. Thus security should prevent unauthorized access to the network and to the information it contains, and assure data integrity (to prevent their unauthorized modification).

A potential violation of security is treated as a threat. The following classes of threats are regarded as security relevant:

- Accidental threats that exist with no premeditated intent (of no malicious nature) including, for example, network malfunctions, operational blunders and software bugs
- Intentional threats that, if realized, may be considered to be attacks (classical intentional threats are espionage and sabotage)

Another classification distinguishes between:

- Passive threats which neither modify information contained in the network nor change network operation or network states (a passive wiretapping to observe information being transmitted over a data circuit is an example)
- Active threats involving the alteration of information (active wiretapping resulting in modifications, deletions, delays, reorderings, duplications, insertion of information messages and their fragments) as well as of network operation or states (malicious change to the routing tables in PSPDNs by an unauthorized person, for example)

Two different kinds of attacks on secure communication can be identified: one is against the networks themselves; the other uses the network only as a means of unauthorized access to another system.

In PDNs and their equivalents in ISDNs the threats to data storage, retrieval and processing in DTEs are not considered. Damages and removals of network hardware are also excluded. On the other hand, the most dangerous threats in networks are threats concerning control (protocols) and data, that is, all logical parts of the communication system. We cannot list all kinds of attacks in PDNs, but the following list gives a few examples:

- Masquerade (an entity pretends to be a different one)
- Replay (data message is repeated to produce an unauthorized effect)
- Data message alteration
- Denial of service (due to blocking of a network or its part, or to changing protocol parameter values)
- Disclosure of transferred or stored information

The network user and provider should always, before designing a security measure, identify and assess the specific threats. Not all threats are exploitable because the attacker lacks the opportunity to apply them. After identifying the vulnerability of the network the designer should analyse the likelihood of threats, assess their consequences, estimate the cost of attacks, cost out potential countermeasures and finally select an appropriate security mechanism, bearing in mind the corresponding cost–benefit trade-off.

Among a large set of security mechanisms we choose four that seem to be most important in PDNs. These are encipherment, access control, digital signature and traffic padding. The oldest mechanism, sometimes considered unique, is encipherment or encryption. Omitting hieroglyphic codes, electronic encipherment was developed during the Second World War, although the first results were published only in 1949 by Claude Shannon, the father of classical information theory [67].

Encipherment is a transformation (by way of substitutions, transpositions, productions (extensions)) of intelligible information, called *cleartext* or *plaintext*, to produce a *ciphertext*: in other words, data whose semantic content is not available for unauthorized persons. The reverse of encipherment is decipherment. Both operations seem to be accomplished in the hardware which serves data encoding and decoding for the error control function. However, in contrast to coders/decoders, encipher E/decipher D algorithms must be kept secret. If their secrecy is lost an entirely new design must be made, which is not only time-consuming but also expensive.

The way out of the impasse is to employ a key *k* with the algorithm. A key in this sense is a parameter which allows a large class of different cipher operations to be accomplished with one fixed algorithm (a device or a chip). If the key is compromised it can be changed and the cipher function remains in use.

Encipherment algorithms can be reversible or irreversible. The former is either symmetric, in which case knowledge of the secret encipherment key implies knowledge of the decipherment key, and vice versa, or asymmetric, referred to as public key encipherment.

Encipherment is performed either in the stream mode, where the cleartext is continuously enciphered, or in the block mode, operating on each fragment of cleartext separately. Continuous cipher applied at the bit or signal element stream is, for example, the scrambling and spread spectrum technique. Of course, such mechanisms are weak and may be easily damaged.

The well-known block cipher is the data encryption standard (DES). Its history reaches back to the early 1970s. The US National Bureau of Standards issued in May 1973, and again one year later, a call for encipherment schemes fitting certain requirements. In response several proposals were sent in and one algorithm looked promising in meeting the demand. The chosen submission came from IBM under the name "Lucifer" with the 128 bit key. It was further modified (the key was reduced to 56 bits) and finally adopted as a Federal standard in November 1976 under the name DES. DES has also been adopted by ANSI where it is known as the data encryption algorithm (DEA). Ten years later this algorithm was approved as an international standard (ISO 8372).

The DES key consists of a 56-bit variable and 8 bits of odd parity. If the variable part is randomly chosen, no technique other than trying all possible keys (over seventy quadrillion) is available. This is, however, computationally infeasible in a reasonable time. To prove this, suppose a test time of $1\,\mu s$ per key: the time needed for checking one half of the key space where the correct key may be found is on average more than 100 years.

The DES is a symmetric cipher because knowledge of the same secret key k is needed at the sending as well as at the receiving side (Fig. 7.7a). Secret keys, however, have to be transported, distributed and managed at least between communicating stations. This is risky and provisions for doing it are not easy throughout the network, and are always costly. In PDNs there are no secure links (shielded lines), over which secret keys could be transferred. The only solution is to deliver secret keys physically (by car or by mail) or using separate transmission paths outside the network (by radio links, for example).

A surprising revelation appeared in 1976 in a paper by Diffie and Hellman [18] that suggested an asymmetric system based upon two keys: a sender's encipherment key k_E and receiver's decipherment key k_D. While the former is *public*, transported over any insecure channel, the latter must be kept secret. This scheme is shown in Fig. 7.7b, which shows the decrease of the secret area in comparison with the symmetric system in Fig. 7.7a. Both keys are of course derived from the starting key chosen at random. However, knowledge of the public key allows one neither to calculate the starting nor the decipherment (secret) key. The penalty of such a simplification of encipherment key distribution is that no kind of authentication of the received message is provided. Any intruder can generate a false message using the public key and the receiver is not warned that this information comes from an enemy. The lack of authentication can be overcome in several ways: the digital signature (see below) is the strongest solution. Another problem surrounding the public key is the choice of corresponding functions to derive public and secret keys. The first public key cipher system used in practice was the one devised by Rivest, Shamir and Adleman, often abbreviated as RSA. Encipherment and decipherment in the RSA method employ the power function modulo a prime.

Encipherment supports several security services, such as authentication (peer entity, data origin), data confidentiality (connection/connectionless, selective field, traffic flow), data integrity (connection with/without recovery, selective field connection/connectionless) and origin/delivery non-repudiation.

Access control is a very important mechanism, particularly in public networks. PDNs and ISDNs offer the ability to access any DTE (network subscriber) on the network from any other DTE in the world via other networks with interworking units. However, this advantage can make the network vulnerable to interference by competitors or hackers.

Access control mechanisms are based on the use of, for example, authentication information such as password, route of attempted access, time and duration of access, etc. In PSPDNs a password port can be provided at a PAD for start–stop DTEs. A PAD may have built-in password security for both local and remote users which prohibits unauthorized call set-up. At the called DTE, a valid password must be received before any data is passed. The receipt of an incorrect password will cause the call to be cleared with no indication that it was due to an invalid attempt.

The X.25 packet level protocol can provide an invisible security procedure called "calling address validation". This feature is made available through location in the call request packet where the network user address can be inserted automatically just by the calling PAD. PDNs also have a certain security measure among user facilities, for a closed user group for example.

Detection of unauthorized network usage should also be considered. Repeated calls to certain areas may indicate a hacker trying to break a password. Switching nodes can run procedures that can capture such calls and check the calling address to determine their origin and then provide an alarm to higher layers or to a network management body. The purpose of a digital signature is to provide the receiver of an information message with irrefutable assurance that the message was actually sent by the indicated sender. Digital mechanisms are appropriate for the provision of authentication (peer entity and data origin), connectionless integrity and non-repudiation of origin and delivery. Two procedures are defined: signing a data unit and verifying a signed data unit. The former procedure uses

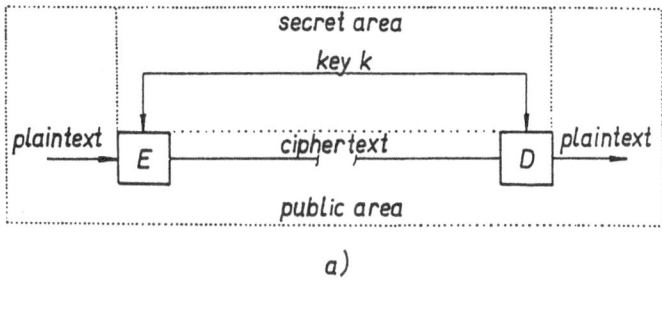

a)

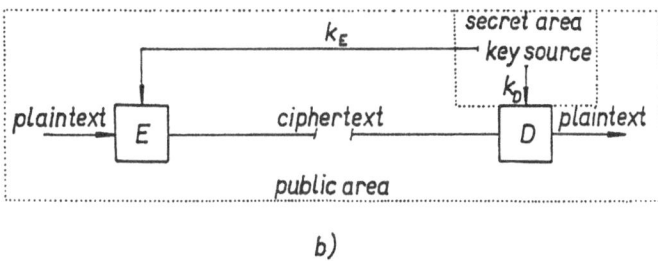

b)

Fig. 7.7. The principle of key cyphers. **a** Symmetric (secret key); **b** asymmetric (public key).

information available publicly, but from which the signer's private information cannot be deduced. The private key encipherment mechanism (as in DES) can be used only between communicating parties who mutually trust each other. Otherwise, if these parties are in dispute, a trustworthy third party (an arbiter) is needed. However, a digital signature can be produced by the public key mechanisms provided that the public and private keys are mutually inverse in a certain sense (as with the RSA scheme). In that case the sender's encipherment key serves the signature and has to be kept secret while the receiver's decipherment key may be publicly distributed and provides for signature verification (see again Fig. 7.7b where the secret zone is at the sender side).

A *digital signature mechanism* is provided in many information systems, primarily in those using a vulnerable public network for information exchange, as in message handling and directory services. In order to conceal from the intruder not only the semantics of messages transmitted but also the knowledge of message flow destination and quantity, *traffic padding* is provided. An enemy tapping the wire is able to recognize the syntax of the data stream even if encrypted (start–stop characters, frames, packets, blocks) and deduce the confidential information. This problem is overcome by encryption of the whole text (including addresses) and by filling out all pauses by encrypted nonsense, which is of course removed only by the authorized addressee (like the zeros inserted in the HDLC frames).

A very important task is to place security mechanisms in the network. Two aspects should be taken in account simultaneously: structural and architectural.

Structural security deals with the problem of where (i.e. in which device) the security mechanism will be implemented. Several possibilities exist:

- DTE security protecting only user devices or some parts of them. Beside physical protection means (several types of terminal locks and screens), personal check-up and education together with personal self-regulation play an important role in secure communication over networks. DTE security is the user's responsibility
- Circuit or link security (Fig. 7.8a), which protects only circuits or links, respectively (subscriber and interexchange as well). In this case network nodes (exchanges, PADs, concentrators) are exposed to threats while the encipherment involves all data (user and control) passing data circuits or data links. In circuit security, the necessary equipment forms part of the DCEs; link security requires this equipment to be shifted up to control units. The network provider is responsible for circuit/link security, in co-operation with the network user at subscriber level
- Network security (Fig. 7.8b), where the entire network, excluding DTEs but including DCEs, is protected by the network provider. Encipherment in PSPDNs may employ different sections between packet switching nodes which increase overall security. The special modules, however, must

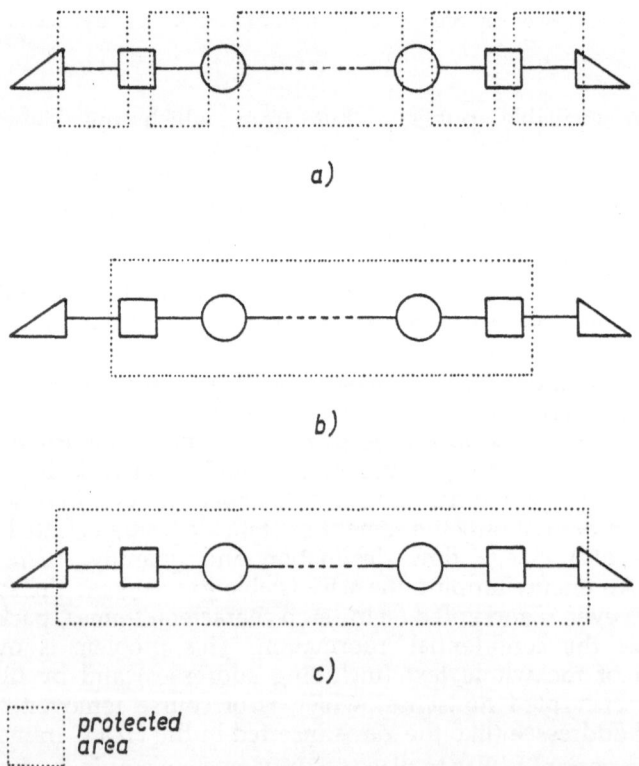

a)

b)

c)

protected area

Fig. 7.8. Possible deployment of security mechanisms in a network. **a** Link security; **b** network security; **c** end-to-end security.

manage all keys corresponding to incoming and outgoing links. In PSPDNs with PAD ports only user data are enciphered at the DTE–PAD interface and control data remain plain in order to control communication during the data exchange phase. The so-called node-by-node encipherment also belongs to network security, and solves the problem of separating user and control data for encipherment

- End-to-end security is the strongest method because it protects the whole chain, DTE–network–DTE (Fig. 7.8c). This type of security involves not only encryption but also access control and digital signature mechanisms and is of course the user's responsibility

From this it follows that no public network is protected, since the network providers have no effective means to guarantee security. This is the reason why certain users prefer private networks to public ones even though they do not want to spend much money on security. This situation should change in the near future through the introduction of efficient network mechanisms which would encourage all users to be satisfied at least at the basic level.

Architectural security deals with the placement of security mechanisms into OSI layers. ISO standard 7498, Part 2 (see Appendix 4) gives a very exhaustive overview and the pros and cons for the security services and mechanisms to be corresponding layer services and functions, respectively. Here we survey only briefly the three lowest layers, with extension to higher layers with respect to public data networks.

Two security services are recommended to be provided at the physical layer. These are connection and traffic flow confidentiality supported principally by encipherment (encryption). Specific forms of bit stream encipherment have already been mentioned (spread spectrum and scrambling).

The data link layer is recommended to be excluded from the placement of any security services. The only exception – connection and connectionless confidentiality – is admitted for certain historical reasons. This is true for PSPDNs if complete data fields in frames are encrypted since they embrace packet headings, that is, all information required to route calls. However, the CSPDN architecture allows link layer transparency and hence end-to-end control. In that case almost all security services recommended for the transport layer can be applied at the link layer, too.

Security services at the network layer are interesting, particularly for relaying, routing and interworking. The following services may be provided (singly or in a combination): peer entity authentication, access control, connection or connectionless confidentiality, traffic flow confidentiality, connectionless integrity and connection integrity without recovery (recovery is not universally available at the network layer). These services are supported by all of the mechanisms mentioned above, to which the prearrangement of routes by the network provider upon demand of the network user, if he detects a manipulation attack (routing control), may be added. Routing control replaces, if possible, encipherment for connection, connectionless and traffic flow confidentiality.

All network security services may also be provided at the transport layer (except traffic flow confidentiality). The security mechanisms may be applied for individual transport connections.

The session layer provides no security service, the presentation layer is acceptable for confidentiality, and finally the application layer is the only layer which is marked as a possible location for all security services. Of course, it is not necessary (or recommended) to apply any security mechanism in more than one layer. The choice of the optimal placement is the network designer's concern, depending upon network user requirements and network provider possibilities. A discussion of the major issues regarding encipherment is including in Annex C of the standardization document, ISO 7498 Part 2.

8 ■ Network Standards [22,33,35,45,52,75]

8.1 The Importance of Standardization for PDNs

A PDN is a very complex system requiring considerable skills in its use. It is operated by a telecommunications administration or a recognized private operating agency (RPOA), as any other public telecommunication network. Its users are organizations, companies, institutions, authorities, enterprises and individuals. The network interfaces a variety of DTEs manufactured by various companies. That all leads to the conclusion that without proper regulation a PDN could not be created, operated, expanded or initiated. The only effective forms of regulation are international standards which are at present respected by manufacturers, operating entities and users.

Unlike these permanently valid basic standardization principles, standardization organizations are undergoing changes in structure as well as in methods of work. This has to be considered when using the facts compiled in this chapter (late 1990).

PDN international standardization corresponds to the standardization of the OSI. An OSI base standard specifies the services provided by, and the protocol associated with, a layer of the OSI model. The base standards are developed mainly by ISO and CCITT (see the following sections). The different approaches of these two bodies are identified by the statement that products should work in accordance with CCITT recommendations or conform to ISO international standards. They are standards of a very general nature. Therefore, in 1983, the concept of functional standards was introduced. The functional standards group together base standards as required for specific applications such as electronic mail. Work on functional standards is done by a group of standards-making bodies in various parts of the world: CEN/CENELEC (European Committee for Standardization/ European Committee for Electrotechnical Standardization) for European products, ETSI (European Telecommunications Standards Institute) for use throughout Europe, NIST (National Institute for Science and Technology) for North America, AOW (Asia Oceania Workshop) for Japan, Korea, China, Australia and the Pacific Rim, EWOS (European Workshop on Open Systems) for use throughout Europe.

Collaboration between these bodies enables harmonization of functional standards as unified input to ISO for the development of international standard profiles (ISPs). ISPs take into account the interests of manufacturers, users, the PTTs and other regional and national authorities. An ISP is elaborated in three phases:

- Elaboration of proposals by a Feeders' Forum including ECMA (European Computer Manufacturers Association) and their formalization by ISO
- Approval by regional and national authorities
- Checking and final approval by ISO

8.2 Worldwide Standardization Organizations
[34,45,51,58]

For PDNs and PDN functions in the ISDN the most relevant rules of interworking are set by the International Telecommunication Union (ITU), which is a UNO organization specializing in telecommunication and radiocommunication. Founded under the name of the International Telegraph Union in 1865, it has attracted PTT administrations and RPOAs from almost all UNO member countries.

The ITU summit is the Conference of Plenipotentiaries, which governs all ITU activities. In the period between conferences the ITU is headed by the Administrative Council convened annually. Permanent ITU entities are the General Secretariat, the International Frequency Registration Board (IFRB), the International Radio Consultative Committee (CCIR) and the International Telegraph and Telephone Consultative Committee (CCITT), whose structure is shown in Fig. 8.1. The most important factor contributing to standardiza-

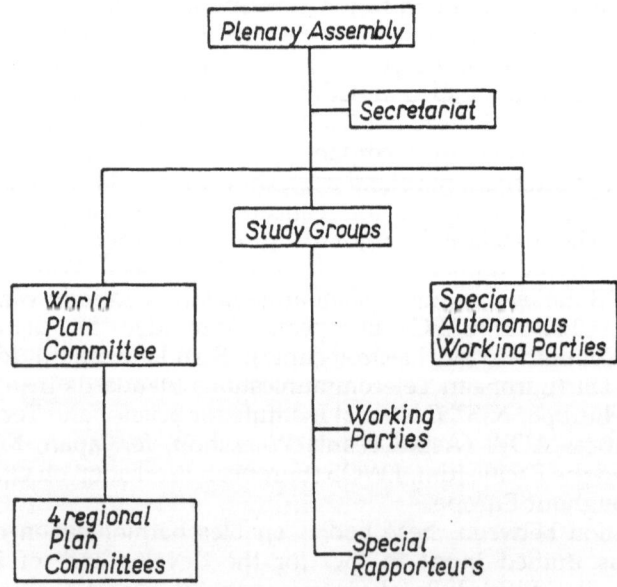

Fig. 8.1. Basic organizational structure of CCITT.

tion of national PDNs and ISDNs with the aim of uniting them into corresponding international networks is the CCITT. It has existed under this name since 1956 when it was formed by the merger of CCIT (International Telegraph Consultative Committee) and CCIF (International Telephone Consultative Committee), which had existed since the 1920s. The bulk of CCITT work is done in study groups (SGs). After the IXth CCITT Plenary Assembly in 1988 there were 15 SGs in total:

SG I   Services
SG II   Network operation
SG III   Tariffs and accounting principles
SG IV   Maintenance
SG V   Protection against electromagnetic effects
SG VI   Outside plant
SG VII   Data communication networks
SG VIII   Terminals for telematic services
SG IX   Telegraph networks and terminal equipment
SG X   Languages for telecommunications applications
SG XI   Switching and signalling
SG XII   Transmission performance of telephone networks and terminals
SG XV   Transmission systems and equipment
SG XVII   Data transmission over the telephone network
SG XVIII   ISDN

The actual work of an SG is done in working parties (WPs). SG VII, responsible for X-series recommendations, had the following WPs in study period 1989–1992:

WP VII/1 Network services, facilities and quality of service
WP VII/2 Network access interfaces
WP VII/3 Interworking, switching and signalling
WP VII/4 Message handling, directory systems and related studies
WP VII/5 Numbering, routing and the layered model

SG VII has full responsibility for PDN standardization. However, other SGs are, to a certain extent, related to PDNs (in particular SGs III, VIII, IX and XVII), and SG XVIII is also responsible for the integration of PDN-type services into the ISDN.

In addition to SGs there are special autonomous groups (GAS) established to elaborate handbooks to aid ITU members in the development of certain telecommunication networks and services. In the study period 1985–1988 GAS II produced a handbook on the strategy for PDNs [22].

Work in the CCITT is planned in four-year periods (study periods). At the end of each period the CCITT Plenary Assembly takes place. It approves new and revised recommendations. Each recommendation gets a letter designation pertinent to the SG which has produced it or to a certain domain of specialization, and a number, the first part of which is characteristic of the section within this domain.

The letter X is allocated to SG VII and the main sections are X.1 to X.15 – services and facilities, X.20 to X.32 – interfaces, X.40 to X.82 – transmission, signalling and switching, X.92 to X.141 – network aspects, X.150 – maintenance, X.180 to X.181 – administrative arrangements, X.200 series –

open systems interconnection, X.300 series – interworking between networks, mobile data transmission systems, X.400 series – message handling systems, X.500 series – directory.

The recommendations are published in books of the same colour covering a certain study period (Plenary Assembly): red (1960, 1984), blue (1964, 1988), white (1968), green (1972), orange (1976) and yellow (1980). A survey of 1988 blue books is given in Appendix 3.

In the past it took, as a rule, one to two study periods to prepare a recommendation, but current needs demand acceleration of the work. Therefore an SG nominates a special *rapporteur* who, with a group of experts, studies the issue and prepares a draft. In this way it is possible to fulfil very complex and difficult tasks. A typical example is packet switching, which in 1970 was an almost unknown concept in CCITT. The special *rapporteur* enlarged the draft during the relatively short period from 1973 to 1976. Another approach to bring recommendations into practice within a shorter term is the so-called accelerated procedure for the provisional approval of recommendations between Plenary Assembly sessions. Definitive adoption of a recommendation is a matter for the following Plenary Assembly.

The Plenary Assembly also approves study questions for the next four-year period. They are the base upon which SGs develop their activities.

CCITT issues recommendations which at first sight could seem less binding than standards. This is not so because CCITT recommendations are used by ITU members as strict standards without their adoption by national standardization bodies.

The most significant organization for worldwide international standardization is the International Organisation for Standardisation (ISO), founded in 1946. Its mission is to promote the development of standards in the world with a view to facilitating the international exchange of goods and services and to develop co-operation in the sphere of intellectual, scientific, technological and economic activity. Membership is by country. Member bodies are national standardization organizations.

The highest policy-making body within the ISO is the General Assembly of Members (Fig. 8.2), which meets every three years. For administrative matters the chief component is the Council which is consulted by its Planning Committee (PLACO). Technical work is carried out within technical committees (TCs) for standardization in particular areas. A technical committee can create subcommittees (SCs) to deal with defined aspects of its work. Both TCs and SCs can create working groups (WGs) for special problems.

The technical committee responsible for data communications is TC 97 Computers and Information Processing which in 1987 became a joint technical committee JTC 1 ISO/IEC. It is subdivided into the following subcommittees:

SC 1   Vocabulary
SC 2   Character sets and coding
SC 6   Telecommunications and information exchange between systems
SC 7   Software development and system documentation
SC 11   Flexible magnetic media for digital data interchange
SC 14   Representation of data elements

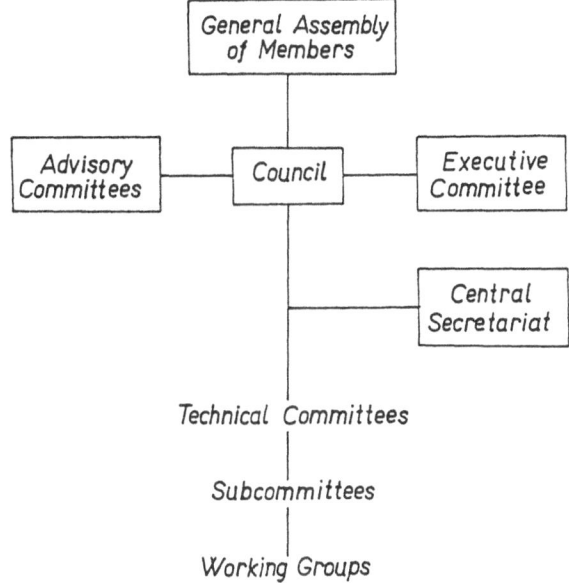

Fig. 8.2. Basic organizational structure of ISO.

SC 15 Labelling and file structure
SC 17 Identification and credit cards
SC 18 Text and office systems
SC 21 Information retrieval, transfer and management for open systems interconnection
SC 22 Languages
SC 23 Optical digital data disk
SC 24 Computer graphics
SC 25 Interconnection of information technology equipment
SC 26 Microprocessor systems
SC 27 Information technology equipment
SC 28 Office machines

Subcommittee SC 6, originally created only for data transmission, deals with the standardization of the first four OSI layers, hence the names of its WGs (by the end of 1991):

WG 1 Data link layer
WG 2 Network layer
WG 3 Physical layer
WG 4 Transport layer
WG 6 Private integrated service networking

The original SC 16, established in March 1977 for the elaboration of open systems interconnection architecture, became SC 21. It deals with the three top layers of the OSI reference model within six WGs (by the end of 1991):

WG 1   OSI architecture
WG 3   Database
WG 4   OSI management
WG 5   Specific application services
WG 6   OSI session, presentation and common application services
WG 7   Basic reference model of open distributed processing services

The technical committee charged with the task of developing a new standard will set a working group to produce a draft proposal. Usually the group is able to find a document to serve as a working draft. This may be a standard already drawn up by some other organization or submitted by a national standardization body. After refining, it is submitted to the parent committee as a draft proposal.

A draft proposal is registered by the ISO Secretariat and designated DP. The number allocated to the DP will stay with that standard from then on. After full agreement of the DP it becomes a draft international standard (DIS).

To be accepted, a DIS must be approved by the majority of P-members (active participants) of the technical committee and by at least 75 per cent of all votes cast. Once a DIS is approved it is submitted to the ISO Council for certification after which it is published as an international standard, designated IS... . Though ISO standards are international standards, countries adopt them as national standards usually without significant modification.

The International Electrotechnical Commission (IEC) formed in 1906 is ISO's electrical division, with standardization activities in the electrical and electronic domain, especially from the point of view of reliability, protection and compatibility. Each country is represented by its National Electrotechnical Committee representative of manufacturers, users, government, teaching and professional bodies.

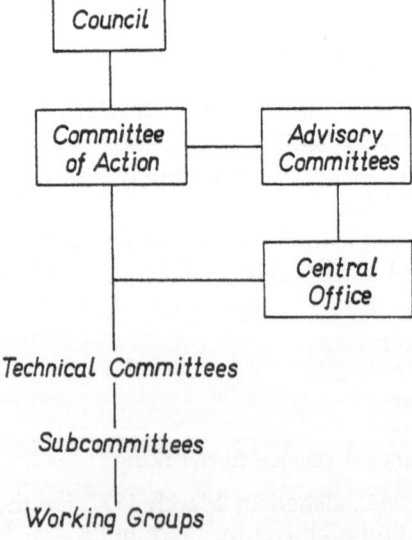

Fig. 8.3. Basic organizational structure of the International Electrotechnical Commission (IEC).

The IEC is headed by the IEC Council (Fig. 8.3). Specific matters are delegated to the Committee of Action. The organization of technical committees, subcommittees and working groups is similar to that of ISO.

Of direct significance for PDNs is the joint technical committee ISO/IEC JTC 1 and TC 74 Safety of Data Processing Equipment and Office Machines (including DTEs, DCEs and other PDN components). Connectors are dealt with in SC 48B, optical local area networks in TC 83.

The International Federation for Information Processing (IFIP) has been created with the aim of improving international co-operation in the information sciences and is a formative body for standards. IFIP congresses are held every three years. Its technical committees and working groups forward the development of computer sciences. TC 6 is responsible for data communications, its WG 6.1 covers the domain of architectures and protocols, WG 6.5 deals with message handling and WG 6.6 is devoted to network management.

8.3 Regional Standardization Organizations [58,73]

The most relevant regions for telecommunications standardization in general, with PDN standardization as its special branch, are Europe, North America and the Far East/Pacific Rim. Telecommunication standardization in Europe is based upon CCITT recommendations, which the European PTT administrations review within the CEPT (European Conference of PTT Administrations). The transition from a CCITT recommendation to a CEPT recommendation and then to a national technical regulation results in national services which are mutually incompatible. Examples of this are videotex, mobile telephony and packet switching.

In 1984 the Commission of the European Communities, in consultation with the Senior Officials Group on Telecommunications (SOGT), entrusted CEPT to prepare common technical specifications for the approval of telecommunication terminals and the purchase of network equipment. During the following years a large number of recommendations were produced. Unfortunately, this procedure has not been as successful as was hoped, because CEPT, which lacks an efficient structure of guaranteed resources, has proved unable to produce the necessary standards within the deadlines required.

In June 1987 CEPT published the "Green Paper on the development of common market in telecommunication services and equipment". Among the Action Lines proposed, special attention was given to the problems of the development of telecommunication standards and the creation of an institute for this purpose. In response to this, CEPT convened a meeting of the PTT Directors-General in September 1987 to establish such an institute. In March 1988 the first General Assembly of the European Telecommunications Standards Institute (ETSI) declared that ETSI was a reality. Its members represent administrations, operators, manufacturers, research bodies and users. ETSI has been given responsibility for the quick establishment of European-wide technical standards for telecommunications and the related areas of broadcasting and office information technologies. ETSI products are:

ETSs (European Telecommunication Standards) or I-ETSs (interim ETSs needing further development) for CEPT members. Some of them are candidates to NETs (Normes européennes de télécommunications – European telecommunication standards) which guarantee the common conformity specifications for type approval of terminal equipments to be connected to public networks.

Examples of approved ETSI standards related to PSPDNs and the ISDN are:

ETS 300 052   ISDN: Multiple subscriber number supplementary service, digital subscriber signalling system No. 1 protocol

ETS 300 055   ISDN: Terminal portability supplementary service, digital subscriber signalling system No. 1 protocol

ETS 300 058   ISDN: Call waiting supplementary service, digital subscriber signalling system No. 1 protocol

ETS 300 061   ISDN: Subaddressing supplementary service, digital subscriber signalling system No. 1 protocol

ETS 300 064   ISDN: Direct dialling in supplementary service, digital subscriber signalling system No. 1 protocol

ETS 300 102–1 ISDN: User-network interface layer 3, specifications for basic call control

ETS 300 102–2 ISDN: User-network interface layer 3, specifications for basic call control, SDL diagrams

ETS 300 103   ISDN: Support of X.21, X.21bis and X.20bis DTEs by an ISDN, synchronous and asynchronous terminal adaption functions

ETS 300 104   ISDN: Attachment requirements for TEs to connect to an ISDN using ISDN basic access, layer 3 aspects (candidate to NET 3)

ETS 300 123   Attachment requirements for DTE to connect to PSPDN using CCITT recommendation X.25 (1984) interface, requirements applicable to DTEs subscribing to modulo 128 operation (candidate to NET 2)

ETS 300 124   Attachment requirements for DTE to connect to PSPDN using CCITT recommendation X.25 (1984) interface, requirements applicable to DTEs subscribing to multilink operation (candidate to NET 2)

ETS 300 125   ISDN: User-network interface data link layer specification, application of CCITT recommendations Q.920/I.440 and Q.921/I.441

As in other international standardization organizations, most of the actual elaboration of standards is done in technical committees, often subdivided into sub-technical committees (Fig. 8.4). Besides these, there are project teams of invited experts for the preparation of draft standards. The work is scheduled by a detailed work programme. ETSI has the following TCs (or non-technical committees with similar organization of work):

NA   Network Aspects
BT   Business Telecommunications
SPS   Signalling, Protocols and Switching
TM   Transmission and Multiplexing

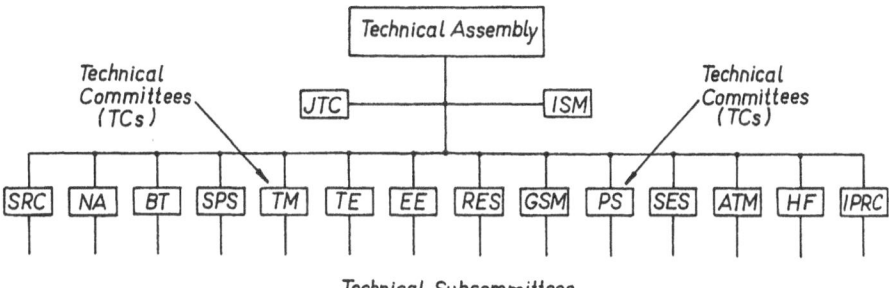

Technical Subcommittees

Fig. 8.4. Basic structure of the European Telecommunications Standards Institute (ETSI).

TE   Terminal Equipment
GSM Special Mobile Group
PS   Paging Systems
SES   Satellite Earth Stations
EE   Equipment Engineering
ATM Advanced Testing Methods
HF   Human Factors
RES   Radio Equipment and Systems
IPRC Intellectual Property Rights Committee
SRC Strategic Review Committee on Telecommunications Terminal
      Equipment
ISM ISDN Standards Management Special Committee
JTC Joint EBU/ETSI Technical Committee

The ECMA (European Computer Manufacturers Association) was founded in 1960 to protect the interests of European computer manufacturers by preparing relevant draft standards and recommendations and by promoting their acceptance, primarily by ISO and CCITT. The Plenary Assembly of ECMA convenes twice a year, to enable a prompt response to the development of computer technology. In this way ECMA contributed to early standardization of OSI protocols and LAN profiles. All this work is performed in technical committees (TCs) and their subordinated task groups (TGs). From the point of view of PDNs the TCs of greatest significance are TC 1 Input and Output Codes and TC 32 Communication, Networks and Systems Interconnection, with TG 1 Data Networks, TG 3 Local Area Networks, TG 4 OSI Control and TG 7 Transport and Network Layer.

In the development of US standards for telecommunications and informatics two US standards committees play an important role: the American National Standards Institute (ANSI) Accredited Standards Committees (ASCs) X3 and T1. The US Organization for the CCITT (USCCITT) co-ordinates US participation in the CCITT. The Electronic Industries Association (EIA) and the Institute of Electrical and Electronics Engineers (IEEE) are also involved in telecommunication standards activities.

ASC X3 Information Processing Systems evolved from a 1980 merger of the ANSI Committees X3, Information Systems and X4 Office Machines. The secretariat for X3 is the Computer and Business Equipment Manufacturers

Association (CBEMA). The scope of X3 covers standardization in the areas of computers, information processing, peripheral equipment and the related devices and media, and that of functional characteristics, particularly in those areas that influence their operation.

ASC T1 develops standards and technical reports related to interfaces for US networks of the North American telecommunications system. The secretariat for T1 is the Exchange Carriers Standards Association (ECSA). T1 standardization concerns the interconnection and interoperability of telecommunications networks at interfaces with end user systems, carriers, and information and enhanced service providers. Interest is focused on switching, signalling, transmission, performance, operation and maintenance procedures at points of interconnections and provisioning methods and documentation.

The Electronics Industries Association (EIA) is a trade association for US manufacturers of electronic equipment and components. As a standards provider in electronics, it has become more involved in telecommunications and fibre optics. Standards developed by EIA may be submitted to ANSI as proposed American National Standards.

The Institute of Electrical and Electronics Engineers (IEEE) has a membership of about 300 000 technical experts and is a certain source of standards in electronics and computers. Standards developed by the IEEE may be published by the IEEE or submitted to ANSI as proposed American National Standards: a series of LAN standards, for example.

Japanese standardization activities in the field of telecommunications have a major influence upon telecommunications in the Far East and Pacific Rim. The activities are concentrated in the Telecommunication Technology Committee (TTC) whose main specialization is standardization in the field of telecommunications and is involved in the following main activities:

- Developing protocols and standards for telecommunication networks
- Conducting studies and research for these protocols and standards
- Disseminating these protocols and standards

Equipment conforming to the standards are given "seals of approval". Part of TTC's activities is evaluation of the effect of standards on the Japanese market and digitized services in the Japanese telecommunications environment.

The TTC is subdivided into four technical subcommittees: network-to-network interfaces, network-to-terminal equipment interfaces, higher layer, and specialized subjects (including PBXs and LANs). They are further subdivided into working groups (WGs) entrusted with specialized items, such as interface protocol ISDN/non-ISDN, common higher level, telematic terminals, study of international trends in protocols and standards.

9 ■ Trends in Networks and Services [24]

Any prediction about the future development of PDNs and the services which they support should be based upon knowledge of utilizable technology and on an analysis of users' needs. The issue needs to be considered in association with all the other telecommunication networks and services, including private systems for limited areas (LANs and their interconnection into WANs, for example) and limited user populations (private networks and closed user groups).

The pace of development is given by advances in technology and by factors influencing the development of human society and national economy. The leading edge of progress in telecommunications is telecommunication research, much of which is accomplished by manufacturers of telecommunication equipment as well as by operating organizations (PTT administrations and RPOAs). Interworking of telecommunication networks on all connectivity levels, from local to national to global, is supported by corresponding levels of standardization.

As to the technological base for the development of PDNs and PDN supported services, two factors are the most important: the availability of suitable transmission media (optical fibre and satellite links being the most important ones) and the application of advanced microelectronics and computer techniques. The milestones in the development of modern telecommunication networks can be expressed by two concepts: digitization and integration. Digitization has paved the way to integration of telecommunication services within a common transmission and switching system and to the integration of transmission and switching techniques within one exchange on the basis of advanced TDM. This system enables the creation of a very flexible digital transmission network whose function is equivalent to that of automatic or distant controlled operation of an exchange's distribution frame.

Another enhancement of PDNs and PDN services consists of the perfecting of high-speed packet switching, asynchronous transfer mode. The remarkable advantages of this technique make it a candidate for utilization in other telecommunication disciplines.

The changing telecommunication environment will probably have an effect on the existence of dedicated PDNs, which will survive in the neighbourhood of ISDNs only if this survival is technically and economically viable. The survival of data transmission and its related value added services (especially message handling) can be expected in new advanced telecommunication networks, *intelligent networks* and networks oriented towards interpersonal communication, for example.

Appendix 1. Basic Terms

The following terms have been selected for the purpose of this book only. They should serve for better understanding but they do not claim to be exhaustive. Their explanations, rather than definitions, are as simplified as possible and, though the ISO and CCITT definitions have been taken into account, they are mostly due to the authors. Terms are presented in alphabetical order, rather than grouped or logically arranged.

Access protocol: Protocol adopted at an interface between a user and a network to enable the user to employ the services of that network

Address: An unambiguous name referring to the service access point at which the OSI service is made available to the service user by the service provider

Addressing domain: Set of addresses that are assignable to objects of a particular type

Anisochronous data network: Data network providing anisochronous transmission

Anisochronous transmission: A transmission process such that between any two instants of transition from one signal condition (e.g. condition "1" or "0" in a binary signal) to another transition in the same group (frame or character) there is always an integral number of unit intervals (shortest theoretical duration between two transitions); between two such instants located in different groups, there is not always an integral number of unit intervals

Availability: Ability of a system or network to be in a state to perform a required function or to provide a required service at any instant of time within a given interval

Bearer channel: Common channel of a multichannel system

Bearer service: *See* Transmission service

Bidirectional: A qualification which implies that the transmission of information occurs in both directions

Block: Transport layer protocol data unit

Byte: Group of usually 8 or 9 bits operated upon as a unit

CCITT service: Telecommunication service offered by PTT administrations or RPOAs in compliance with CCITT recommendations

Circuit switched public data network (CSPDN): Public data network providing data transmission over a chain of physical circuits constituting a connection

Circuit switching: Interconnection of circuits for the exclusive use of a connection for the duration of a communication

Closed user group: A facility assigned to specific users of a public data transmission service(s) which permits such users to communicate with each other but precludes communication with all other users of the service or services

Code transparency: Independence for correct functioning upon the character set or code used

Communication: Transfer of any kind of information controlled by protocols according to agreed conventions

Communication service: Transmission service with a portion of data processing with regard to its users

Connection: Temporary association of switching and transmission functional units set up to provide a means of information transfer between two or more systems

Connection-mode: Mode of information transfer over a pre-established connection

Connectionless-mode: Mode of information transfer in self-contained protocol data units without establishing a connection

Data: A presentation of facts, concepts, or instructions in a formalized manner suitable for communication, interpretation, or processing by automatic means

Data channel: Means of unidirectional transmission of data-carrying signals

Data circuit: Means of bidirectional transmission of data signals

Data circuit terminating equipment (DCE): In the PDN a facility designated for interworking with a corresponding data terminal equipment (DTE) via an appropriate interface

Data communication: Transfer of data controlled by protocols according to agreed conventions

Data link: Means of data communication between entities comprising data sources and data sinks

Data network: Network designated for data communication

Data security: Ability of a system or network to prevent disclosure, transfer, modification or destruction, whether accidental or intentional

Data signal: A time-dependent value attached to a physical phenomenon and conveying data

Data signalling rate: Total number of bits transferred over a data channel during a second. The equivalent term "bit rate" is used for digital transmission (not only for data transmission)

Data terminal equipment (DTE): Equipment comprising a data source and/or a data sink, control unit, designated for data communication over a data circuit or a data network. The DCE directly co-operates with a DTE

Data transmission: Transfer of data from one place to another by means of data signals over data circuits

Data transmission service: Telecommunication service for data transmission

Datagram service: Transmission service provided by the connectionless-mode

Default value: Value given to any parameter by the network in the absence of a specific value required by the user

Digital network: Network based upon digital signal transmission

Digital signal: Signal consisting of signal elements corresponding to digits of a number system

DTE address: An unambiguous name referring to the DTE–DCE interface in a PDN

Envelope: An agreed arrangement of overhead bits (e.g. for control, identification and stuffing) in the protocol data unit

Format: An agreed arrangement of data, commands and identifiers in protocol data units

Fragment: A piece of user data of limited length

Frame: Link layer protocol data unit

Frequency division multiplex (FDM): a system of sharing a common channel by assigning frequency sub-bands to its tributary channels

Function: Aimed control activity performed by an element or a complex of elements within a communication system or network

Integrated digital network (IDN): Network in which connections established by digital switching are used for the transmission of digital signals

Integrated services digital network (ISDN): Integrated digital network supporting a range of different telecommunication services

Intelligent network: Network reacting to inputs from its environment in an optimum way, as a human being would react

Interface: Common boundary between two associated systems or devices

International Alphabet No. 5 (IA5): Seven-bit coded character set defined by CCITT and ISO (ISO-7)

Interworking function: Functional elements (entities) that intervene in harmonizing the control between two networks (identical or different) to provide transparent data communication

Interworking unit: Unit facilitating the communication between users of two different networks

Intranetwork protocol: Protocol governing transmission or communication between switching nodes within a network

Isochronous signal: A signal where between any two instants of transition from one signal condition (e.g. condition "1" or "0" in a binary signal) to another transition there is always an integral number of unit intervals (shortest theoretical duration between two transitions)

Layer: Result of vertical partitioning of the open systems interconnection defined by entities, functions, services, protocol and service data units, service access points and protocols

Leased circuit: A circuit made available to a user for his/her exclusive use

Local area network (LAN): Private data network laid out within a limited area

Logical channel: Numbered time division channel allocated in particular packets

Message: An ordered series of bits, characters, octets and bytes, intended to convey information

Multiplex: System of sharing a common channel

Multipoint connection: Connection providing more than two end points

Network: A system of switching nodes and transmission circuits that provides transmission or communication to users (subscribers) in accordance with certain rules

Network architecture: Structure and organization of control in a network

Network control centre (NCC): Network functional unit which performs functions of network management, traffic analysis, charging, diagnosis, maintenance, statistical reporting, etc.

Network management: Facilities to control, co-ordinate and monitor the resources that allow communication to take place within the network

Network numbering plan: Addressing domain administered by PTTs or RPOAs

Network performance (NP): Ability of a network to provide services

Network termination (NT): ISDN equipment that performs functions necessary for the operation of access protocols by the network

Octet: Eight-bit byte (group of eight bits) usually having a signification (letter, number, command, identifier)

One-way: A qualification applying to the transfer of information in one direction

Open system: Representation within a generalized, abstract model of those aspects of a real open system that are pertinent to OSI

OSI reference model: Seven-layer model defined in ISO 7498 and CCITT X.200 that describes the general principles of the open systems interconnection

OSI service: A capability of a layer and the layers beneath it within the OSI reference model which is provided to the adjacent higher layer

Packet: Network layer protocol data unit in the packet switched environment

Packet assembler/disassembler (PAD): Functional unit within the PSPDN which enables non-packet-mode DTEs to access that PSPDN

Packet mode DTE (PDTE): A data terminal equipment designated for data transmission in the form of packets

Packet switched network: A network designated for enabling its users to communicate using transmission in the form of packets

Packet switched public data network (PSPDN): A PDN designated for enabling DTEs to communicate using data transmission in the form of packets

Packet switched virtual circuit bearer service: Data transmission service consisting of providing data transmission by means of packets on the user-to-network interface

Packet switching: Constitution of an association between communicating DTEs by means of storing and forwarding packets

Permanent virtual circuit (PVC): Virtual circuit established within the PSPDN for an agreed period of time between DTEs

Point-to-point connection: A connection providing only two end points

Port: Addressable functional unit enabling access to a public data network

Primitive: An abstract, implementation-independent interaction between a service user and a service provider

Private: Established, owned and operated by an individual or organization

Protocol: Set of formats and rules, semantic and syntactic, to perform functions

Protocol data unit: Formatted unit of data exchanged under the control of a protocol

Public: Free for use by anybody who fulfils certain conditions for the usage

Quality of service (QOS): Collective effect of service performances which determines the degree of satisfaction of a user of the service

Real open system: A real system which complies with the requirements of OSI standards in its communication with other real systems

Residual error rate (RER): The ratio of total incorrect, lost, extra or misdelivered user data units (bits, octets, characters, frames, packets, etc.) to total user data units transmitted or received after performing an error control function

Segment: Unit for charging user data volume transmitted over public networks (in PSPDNs 1 S = 64 octets = 512 bits)

Service availability: Ratio of aggregate time during which satisfactory service is, or could be, provided to the total observation period

Signal element: Elementary part of a signal distinguished from others by its nature, magnitude and relative position

Signalling: Exchange of network layer protocol data units in circuit switched environment

Speed transparency: Independence for correct functioning upon signalling rate

Start–stop data terminal equipment (ADTE): DTE capable of transmitting and receiving start–stop signals

Start–stop signal: Anisochronous signal with one signal element for preparing the receiver for reception of a character and one signal element for informing the receiver about the end of transmission of that character

Subscriber circuit: A circuit for the transmission of signals between a station located at the user's premises and the pertaining network exchange

Switched virtual circuit (SVC) or **virtual circuit (VC)**: Virtual circuit established for the duration of communication between two DTEs

Synchronous: Having a fixed time relation between the functioning of a transmitter and that of the corresponding receiver

Telecommunication service: Capability provided by a telecommunication system or network to a system user or network subscriber, respectively

Telematic service: Any CCITT service except telegraph, telephone and data service

Teleservice: *See* Communication service

Terminal adaptor (TA): Functional unit in the ISDN that provides functions necessary for the operation of the access protocols by the user

Terminal equipment (TE): Equipment in the ISDN that performs functions necessary for the operation of the access protocols by the user

Throughput: Mean number of successfully (without errors) transferred user data units (bits, octets, characters, frames etc.) per unit of time

Time division channel: Channel sharing a common path and allocated to a particular, periodically repeated, time slot

Time division multiplex (TDM): A system of sharing a common channel (bearer channel) by allocating time slots to tributary channels

Time-out: Parameter related to an enforced event designated to occur at the expiry of a predetermined period of time

Transit delay: Value of elapsed time between the request for transfer of a protocol data unit and its successful receipt

Transmission service: *See* Bearer service

Two-way alternate (TWA): A qualification applying to the transfer of information in both directions, but only in one direction at a time

Two-way simultaneous (TWS): A qualification applying to the transfer of information in both directions simultaneously

Unidirectional: A qualification which implies that the transmission of information always occurs in one direction

User class of service: A group of users to whom the same kind of service is provided (e.g. defined by transmission mode, data signalling rate and method of switching)

User facility: A supplementary service, the provision of which the user asks for without time limitation, for an agreed period of time, or on a per-call basis

Virtual network: Network providing the packet switched virtual bearer service

Window size: Number of protocol data units that are allowed to be transmitted before an acknowledgement is received.

Appendix 2. Network Identification Codes

This appendix highlights the structure of the addressing system in PDNs and ISDNs, and indicates the names or acronyms of significant existing PDNs. Fig. A2.1 indicates the global numbering of zones, and Fig. A2.2 gives the data network identification codes and numbers of ISDNs in Europe. Table A2.1 explains the allocation of PDN and ISDN identification codes on a worldwide selection of countries and PDNs.

For newer examples of PDNs, the reader is referred to the periodical *ITU Operational Bulletin* issued by the ITU general secretariat in Geneva.

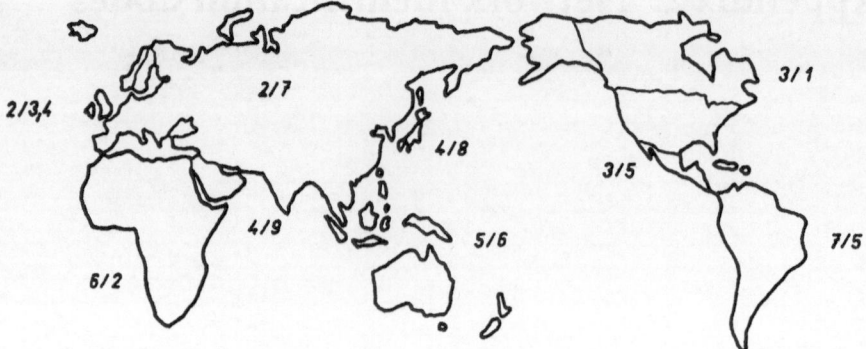

Fig. A2.1. Global numbering of zones.

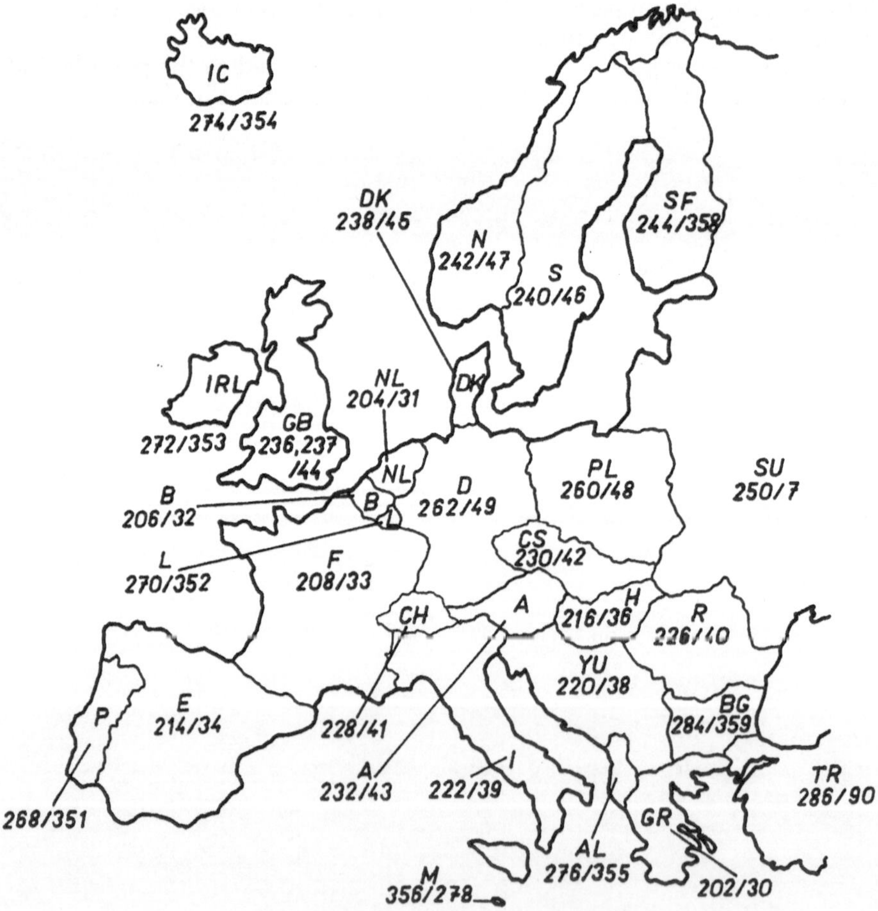

IC
274/354

DK
238/45

N
242/47

SF
244/358

S
240/46

IRL

NL
204/31

DK

GB
272/353   236,237
            /44

NL

B

D
262/49

PL
260/48

SU
250/7

B
206/32

L
270/352

F
208/33

CS
230/42

CH

A

H
216/36

R
226/40

YU
220/38

E
214/34

P

228/41

A
232/43

I
222/39

BG
284/359

TR
286/90

268/351

GR

AL
276/355

202/30

M
356/278

Fig. A2.2. Data network identification code numbers of ISDNs in Europe.

Table A2.1. Data network identification codes

Zone	Country	Network identification codes		Names of selected PDNs
		PDN	ISDN	
Europe	Austria	232	43	DATEX-L, DATEX-P, RADAUS
	Belgium	206	32	DCS, DCS.FAX/DCS.BULK
	Bulgaria	284	359	BULPAK
	Cyprus	280	357	CYTAPAC
	Czechoslovakia	230	42	Eurotel
	Denmark	238	45	DATAPAK, DATEX, DAXNET
	Finland	244	358	DATAPAK, DATEX, DIGIPAK
	France	208	33	TRANSPAC, NTI
	Germany	262	49	DATEX-L, DATEX-P
	Greece	202	30	HELPAK
	Hungary	216	36	(CSDS) NEDIX, PSDS
	Iceland	274	354	ISPAK/ICEPAC
	Ireland	272	353	EIRPAC
	Italy	222	39	ITAPAC
	Luxembourg	270	352	LUXPAC
	Malta	278	356	MALTAPAC
	Netherlands	204	31	DABAS, DATANET 1
	Norway	242	47	DATAPAK, DATEX
	Portugal	268	351	TELEPAC
	Spain	214	34	IBERPAC
	Switzerland	228	41	TELEPAC
	Sweden	240	46	DATAPAK1, DATAPAK2, DATEX
	United Kingdom	236 ⎱ 237 ⎰	44	IPSS, JAIS, MERCURY, PSS PAKNET
	USSR	250	7	NCADE
North America	Canada	302	1	ANICOM, CNCP (CSPDN), CNCP (PSPDN), DATAPAC, GLOBEDAT-C, GLOBEDAT-P, INFOSWITCH
	USA	310 ⎱ to ⎰ 316	1	ACCUNET, ALASCOM/ ALASKANET, AS BELL ATLANTIC, AT&T/ACCUNET, BT TYMNET, FTCC, FTCC-PST, GLOBENET, GTE-HAWAI, ICSS/IPSS, INFONET, ITT-DATEL, ITT-UDTS, JAIS USE-NET, MICROLINK II, NCC-A VAN, PPSS/ US West, PTN-1, RCA-DATEL, TELENET, TRT-ICSS, TRT-IPSS, UNINET, WUI
	Mexico	334	52	TELEPAC
Asia	China	460	86	PKTELCOM
	Hong Kong	454	852	
	India	404	91	GPSS, VIKRAM
	Israel	425	972	ISRANET
	Japan	440 ⎱ 441 ⎰	81	DDX-P, FENIX, GLOBALNET, INTEC Tri-P, JAIS-NET, JENSNET, KINOCOSMONET, MASTERNET, MARUNET, NECC & C/PK-VAN, NIS/TYMNET, NTT DDX, VENUS-C, VENUS-P

Table A2.1.   *(Continued)*

| Zone | Country | Network identification codes | | Names of selected PDNs |
		PDN	ISDN	
	Korea	450	82	DACOM
	Lebanon	415	961	
	Pakistan	410	92	
	Saudi Arabia	420	966	
	Sri Lanka	413	94	
	Syria	417	963	
	United Arab Emirates	424	971	
	Yemen	421	967	
Australia and Oceania	Australia	505	61	AUSTPAC, KEYLINK 2, KEYLINK 7, OTC Data Access
	Indonesia	510	62	SKDP
	Malaysia	502	60	MAYPAC
	New Zealand	530	64	PACNET
	Philippines	515	63	CAPWIRE, PHILCOM, GMCR, ETPI
	Singapore	525	65	TELEPAC
	Thailand	520	66	
Africa	Algeria	603	21	
	Egypt	602	20	
	Ethiopia	636	251	
	Libya	606	21	
	Morocco	604	21	
	Nigeria	621	234	
	Reunion	647	262	
	Senegal	608	221	SENPAC
	South Africa	655	27	
	Sudan	634	249	
	Tanzania	640	255	
	Tunisia	605	21	
	Zimbabwe	648	263	
South America	Argentina	722	54	AGANET, ARPAC
	Bolivia	736	591	
	Brazil	724	55	INTERDATA
	Chile	730	56	ENTEL
	Colombia	732	57	
	Paraguay	744	595	
	Uruguay	748	598	URUPAC
	Venezuela	734	58	

Appendix 3. CCITT Recommendations

This appendix gives a selection of CCITT recommendations of the Blue Books adopted by the CCITT IXth Plenary Assembly in Melbourne, 14–25 November 1988, and edited by the International Telecommunication Union (ITU) in Geneva, 1989. Only those recommendations are listed which are related to the subjects dealt with in this book. The allocation of the recommendations to parts of the Blue Book and to CCITT Study Groups is given in Table A3.1.

Note: Where the heading is not given, the text of the recommendation is to be found under the indicated alternative number.

A.20   Collaboration with other international organizations over data transmission

A.21   Collaboration with other CCITT defined telematic services

D.8   Special conditions for the lease of international end-to-end digital circuits for private service

D.10   General tariff principles for international public data communication services

D.ll   Special tariff principles for international packet switched data communication services by means of the virtual call facility

D.12   Measurement unit for charging by volume in the international data communication service

D.13   Guiding principles to govern the apportioning of accounting rates in international packet switched public data communication relations

D.15   General charging and accounting principles for non-voice services provided by interworking between PDNs

D.20   Special tariff principles for the international circuit switched public data communication services

D.21   Special tariff principles for short transaction transmissions on the international packet switched data networks using the fast select facility with restriction

D.30   Implementation of reverse charging on international public data communication services

D.210 General charging and accounting principles for international telecommunication services provided over the ISDN

D.211 International accounting for the use of the signal transfer point (STP) in CCITT signalling system No.7

Table A3.1. Allocation of PDN-relevant CCITT recommendation series

Recommendation series	Name of series	Fascicle in Blue Book	Study group
A	Organization of working procedures	I.2	
D	Charging and accounting in international services	II.1	III
E	Telephone network and ISDN – Operation, and numbering, quality of service, management, traffic engineering	II.2,3	II
F	Telecommunication services	II.4-6	I
G	Transmission technique	III.1-5	XII, XV, XVIII
M	Maintenance	IV.1,2	IV
Q	Telephone switching and signalling	VI.1-14	XI
R	Telegraph transmission	VII.1	IX
S	Telegraph services terminal equipment	VII.1	IX
T	Terminal equipment and protocols for telematic services	VII.3-7	VIII
U	Telegraph switching	VII.2	IX
V	Data communications over the telephone network	VIII.1	XVIII
X	Data communication networks	VIII.2-8	VII

D.220 Charging and accounting principles to be applied to international circuit mode demand bearer services provided over the ISDN

D.250 General charging and accounting principles for non-voice services provided by interworking between the ISDN and existing data networks

E.163 Numbering plan for the international telephone service

E.164 Numbering plan for the ISDN era

E.166 Numbering plan for interworking in the ISDN era

E.167 ISDN network identification codes

F.50 International public telemessage service

F.160 General operational provisions for the public facsimile services

F.162 Operational provisions for the international store-and-forward facsimile switching service (Comfax)

F.170 Operational provisions for the international public facsimile service between public bureaux (bureaufax)

F.180 Operational provisions for the international public facsimile service between subscribers' stations (telefax)

F.200 Teletex service

F.300 Videotex service

F.350 Application of series T recommendations

F.351 General principles on the presentation of terminal identification to users of the telematic services

F.353 Provision of telematic and data transmission services on the ISDN
F.400 Message handling system and service overview
F.401 Message handling services: naming and addressing for public message handling services
F.410 Message handling services: the public message transfer service
F.415 Message handling services: intercommunication with public physical delivery services
F.420 Message handling services: the public IPM (interpersonal messaging) service
F.421 Message handling services: intercommunication between the IPM services and the Telex service
F.422 Message handling services: intercommunication between the IPM service and the teletex service
F.500 International public directory services
F.600 Service and operational principles for public data transmission services
F.601 Service and operational principles for packet switched data networks
F.710 Teleconference service
G.703 Physical/electrical characteristics of hierarchical digital interfaces
G.704 Synchronous frame structures used at primary and secondary hierarchical levels
I.122 Framework for providing additional packet mode bearer services
I.130 Method for the characterization of telecommunication services supported by an ISDN and network capabilities of an ISDN
I.230 Definition of bearer service categories
I.231 Circuit-mode bearer service categories
I.232 Packet-mode bearer service categories
I.324 ISDN network architecture
I.332 Numbering principles for interworking between ISDNs and dedicated networks with different numbering plans
I.340 ISDN connection types
I.350 General aspects of quality of service and network performance in digital networks, including ISDN
I.351 Recommendations in other series concerning network performance objectives that apply to reference point T of an ISDN
I.352 Network performance objectives for connection processing delays
I.410 General aspects and principles relating to recommendations on ISDN user–network interfaces
I.411 ISDN user–network interfaces: reference configurations
I.412 ISDN user–network interfaces: interface structures and access capabilities
I.420 Basic user–network interface
I.430 Basic user–network interface: layer 1 specification
I.431 Primary rate user–network interface
I.440 (Q.920)
I.441 (Q.921)
I.450 (Q.930)
I.451 (Q.931)
I.452 (Q.932)
I.460 Multiplexing, rate adaption and support of existing interfaces

I.461   (X.30)
I.462   (X.31)
I.463   (V.110)
I.464   Multiplexing, rate adaption and support of existing interfaces for restricted 64 kbit/s transfer capability
I.465   (V.120)
I.470   Relationship of terminal functions to ISDN
I.540   (X.321)
I.550   (X.325)
M.30   Principles for a telecommunications management network
Q.702   Signalling data link (SS No.7)
Q.703   Signalling link (SS No.7)
Q.704   Signalling network functions and messages (SS No.7)
Q.705   Signalling network structure (SS No.7)
Q.706   Message transfer part signalling performance (SS No.7)
Q.709   Hypothetical signalling reference connection (SS No.7)
Q.711   Functional description of the signalling connection control part (SCCP)
Q.712   Definition and function of SCCP messages
Q.713   SCCP formats and codes
Q.714   SCCP procedures
Q.716   SCCP performances
Q.730   ISDN supplementary services ISDN user part (ISUP)
Q.761   Functional description of the ISDN user part of SS No.7.
Q.762   General function of messages and signals
Q.763   Formats and codes
Q.764   Signalling procedures
Q.766   Performance objectives in ISDN applications
Q.771   Functional description of transaction capabilities application part (TCAP)
Q.772   Transaction capabilities information element definitions
Q.773   Transaction capabilities formats and encoding
Q.774   Transaction capabilities
Q.775   Guidelines for using the transaction capabilities application part TCAP)
Q.920   (I.440) ISDN user–network interface data link layer: general aspects
Q.921   (I.441) ISDN user–network interface data link layer specification
Q.921   (I.441) ISDN user–network interface data link layer specification
Q.930   (I.450) ISDN user–network interface layer 3: general aspects
Q.931   (I.451) ISDN user-network interface layer 3 specification for basic call control
Q.932   (I.452) Generic procedures for the control of ISDN supplementary services
R.100   Transmission characteristics of international TDM links
R.101   Code and speed dependent TDM system for anisochronous telegraph and data transmission using bit interleaving
R.102   4800 bit/s code and speed dependent and hybrid TDM systems for anisochronous telegraph and data transmission using bit interleaving
R.111   Code and speed independent TDM system for anisochronous telegraph and data transmission

R.112   TDM hybrid system for anisochronous telegraph and data transmission

S.1   International Telegraph Alphabet No. 2
S.3   Transmission characteristics of the local end and its termination
S.15   Use of the Telex network for data transmission at 50 bauds
S.16   Connection to the Telex network of an automatic terminal using a V.24 DCE–DTE interface
S.18   Conversion between International Telegraph Alphabet No. 2 and International Alphabet No. 5
S.19   Calling and answering in the Telex network with automatic terminal equipment
S.20   Automatic clearing procedure for a Telex terminal
S.31   Transmission characteristics for start–stop data terminal equipment using International Alphabet No. 5

T.1   Standardization of phototelegraph apparatus
T.4   Standardization of group 3 facsimile apparatus for document transmission
T.6   Facsimile coding schemes and coding control functions for group 4 facsimile apparatus
T.50   International Alphabet No. 5
T.51   Coded character set for telematic services
T.60   Terminal equipment for use in the teletex service
T.61   Character repertoire and coded character sets for the international teletex service
T.62   Control procedures for teletex and group 4 facsimile services
T.65   Applicability of telematic protocols and terminal characteristics to computerized communication terminals
T.70   Network-independent basic transport service for the telematic services
T.71   LAPB extended for half-duplex physical level facility
T.72   Terminal capabilities for mixed mode operation
T.73   Document interchange protocol for telematic services
T.90   Characteristics and protocols for terminals for telematic services in ISDN
T.100   International information exchange for interactive videotex
T.101   International interworking for videotex services
T.330   Telematic access to interpersonal message system

V.1   Equivalence between the binary notation symbols and the significant conditions of a two-condition code
V.2   Power levels for data transmission over telephone lines
V.4   General structure of signals of International Alphabet No. 5 code for character oriented data transmission over public data networks
V.5   Standardization of data signalling rates for synchronous data transmission in the general switched telephone network
V.6   Standardization of data signalling rates for synchronous data transmission on leased telephone-type circuits
V.7   Definition of terms concerning data communication over the telephone network
V.10   (X.26) Electrical characteristics for unbalanced double-current interchange circuits for general use with integrated circuit equipment in the field of data communications

V.11   (X.27) Electrical characteristics for balanced double-current interchange circuits for general use with integrated circuit equipment in
the field of data communications

V.13   Simulated error control

V.14   Transmission of start–stop characters over synchronous bearer
channels

V.21   300 bit/s duplex modem standardized for use in the general switched
telephone network

V.22   1200 bit/s duplex modem standardized for use in the general switched
telephone network and on point-to-point two-wire leased telephone-
type circuits

V.22bis 2400 bit/s modem using the frequency division technique standardized for use on the general switched telephone network and on
point-to-point two-wire leased telephone-type circuits

V.23   600/1200 baud modem standardized for use in the general switched
telephone network

V.24   List of definitions for interchange circuits between data terminal
equipment (DTE) and data circuit terminating equipment (DCE)

V.25   Automatic answering equipment and/or parallel automatic calling
equipment in the general switched telephone network including
procedures for disabling of echo control devices for both manually
and automatically established calls

V.25bis Automatic calling and/or answering equipment on the general
switched telephone network (GSTN) using the 100-series interchange
circuits

V.26   2400 bit/s modem standardized for use on four-wire leased
telephone-type circuits

V.26bis 2400/1200 bit/s modem standardized for use in the general switched
telephone network

V.26ter 2400 bit/s duplex modem using the echo cancellation technique
standardized for use on the general switched telephone network and
on point-to-point two-wire leased telephone-type circuits

V.27   4800 bit/s modem with manual equalizer standardized for use on
leased telephone-type circuits

V.27bis 4800/2400 bit/s modem with automatic equalizer standardized for use
on leased telephone-type circuits

V.27ter 4800/2400 bit/s modem standardized for use in the general switched
telephone network

V.28   Electrical characteristics for unbalanced double-current interchange
circuits

V.29   9600 bit/s modem standardized for use on point-to-point four-wire
leased telephone-type circuits

V.31   Electrical characteristics for single-current interchange circuits using
contact closure

V.31bis Electrical characteristics for single-current interchange circuits using
optocouplers

V.32   A family of two-wire duplex modems operating at data signalling
rates of up to 9600 bit/s for use on the general switched telephone
network and on leased telephone-type circuits

V.33   14 400 bit/s modem standardized for use on point-to-point four-wire
leased telephone-type circuits

V.35   Data transmission at 48 kbit/s using 60–108 kHz group band circuits
V.36   Modems for synchronous data transmission using 60–108 kHz group band circuits
V.37   Synchronous data transmission at a data signalling rate higher than 72 kbit/s using 60–108 kHz group band circuits
V.41   Code-independent error-control system
V.42   Error-correcting procedures for DCEs using asynchronous conversion
V.54   Loop test devices for modems
V.100   Interconnection between public data networks (PDNs) and the public switched telephone networks (PSTN)
V.110   (I.463) Support of data DTEs with V-series type interfaces by an integrated services digital network (ISDN)
V.120   (I.465) Support by an ISDN of data terminal equipment with V-series type interfaces with provision for statistical multiplexing
V.230   General data communications interface layer 1 specification
X.1   International user classes of service in public data networks and ISDNs
X.2   International data transmission services and optional user facilities in public data networks and ISDNs
X.3   Packet assembly disassembly facility (PAD) in a PDN
X.4   General structure of signals of International Alphabet No. 5 code for character oriented data transmission over PDNs
X.10   Categories of access for DTE to public data transmission services
X.20   Interface between DTE and DCE for start–stop transmission services on PDNs
X.20bis Use on PDNs of DTE which is designed for interfacing to asynchronous duplex V-series modems
X.21   Interface between DTE and DCE for synchronous operation on PDNs
X.21bis Use on PDNs of DTE which is designed for interfacing to synchronous V-series modems
X.22   Multiplex DTE–DCE interface for user classes 3–6
X.24   List of definitions for interchange circuits between DTE and DCE on PDNs
X.25   Interface between DTE and DCE for terminals operating in the packet mode and connected to PDNs by dedicated circuits
X.26   (V.10)
X.27   (V.11)
X.28   DTE–DCE interface for a start–stop mode DTE accessing the packet assembly/disassembly facility in the same country
X.29   Procedures for the exchange of control information and user data between a packet assembly/disassembly (PAD) and a packet mode DTE or another PAD
X.30   (I.461) Support of X.21, X.21bis and X.20bis based DTEs by an ISDN
X.31   (I.462) Support of a packet mode terminal equipment by an ISDN
X.32   Interface between DTE and DCE for terminals operating in the packet mode of accessing a packet switched PDN through a public switched telephone network or an ISDN or a circuit switched PDN
X.50   Fundamental parameters of a multiplexing scheme for the international interface between synchronous data networks

X.50bis Fundamental parameters of a 48 kbits/s user data signalling rate transmission scheme for the international interface between synchronous data networks

X.51   Fundamental parameters of a multiplexing scheme for the international interface between synchronous data networks using 10-bit envelope structure

X.51bis Fundamental parameters of a 48 kbit/s user data signalling rate transmission scheme for the international interface between synchronous data networks using 10-bit envelope structure

X.52   Method of encoding anisochronous signals into a synchronous bearer

X.53   Numbering of channels on international multiplex links at 64 kbit/s

X.54   Allocation of channels on international multiplex links at 64 kbit/s

X.55   Interface between synchronous data networks using a 6 + 2 envelope structure and single channel per carrier (SCPC) satellite channels

X.56   Interface between synchronous data networks using an 8 + 2 envelope structure and single channel per carrier (SCPC) satellite channels

X.57   Method of transmitting a single lower speed channel on a 64 kbit/s stream

X.58   Fundamental parameters of a multiplexing scheme for the international interface between synchronous non-switched data networks using no envelope structure

X.60   Common channel signalling for circuit switched data applications

X.61   Signalling system No.7: data user part

X.70   Terminal and transit control signalling system for start--stop services on international circuits between anisochronous data networks

X.71   Decentralized terminal and transit control signalling system on international circuits between synchronous data networks

X.75   Packet switched signalling system between public networks providing data transmission services

X.80   Interworking of interexchange signalling systems for circuit switched data services

X.81   Interworking between an ISDN and a circuit switched PDN

X.82   Detailed arrangements for interworking between CSPDNs and PSPDNs based on recommendation T.70

X.92   Hypothetical reference connections for public synchronous data networks

X.96   Call progress signals in PDNs

X.110   International routing principles and routing plan for PDNs

X.121   International numbering plan for PDNs

X.122   Numbering plan between a packet switched PDN (PSPDN) and an ISDN or public switched telephone network (PSTN) in the short term

X.130   Call processing delays in PDNs when providing international synchronous circuit switched data services

X.131   Call blocking in PDNs when providing international synchronous circuit switched data services

X.134   Portion boundaries and packet layer reference events: basis for defining packet switched performance parameters

X.135   Speed of service (delay and throughput) performance values for PDNs when providing international packet switched services

X.136 Accuracy and dependability performance values for PDNs when providing international packet switched services

X.137 Availability performance values for public data networks when providing international packet switched services

X.140 General quality of service parameters for communication via PDNs

X.141 General principles for the detection and correction of errors in PDNs

X.150 Principles of maintenance testing for PDNs using DTE and DCE test loops

X.180 Administrative arrangements for international closed user groups (CUGs)

X.181 Administrative arrangements for the provision of international permanent virtual circuits (PVCs)

X.200 Reference model of open systems interconnection for CCITT applications

X.208 Specification of abstract syntax notation one (ASN.1)

X.209 Specification of basic encoding rules for abstract syntax notation one (ASN.1)

X.210 Open systems interconnection layer service definition conventions

X.211 Physical service definition of open systems interconnection for CCITT applications

X.212 Data link service definition for open systems interconnection for CCITT applications

X.213 Network service definition for open systems interconnection for CCITT applications

X.214 Transport service definition for open systems interconnection for CCITT applications

X.215 Session service definition for open systems interconnection for CCITT applications

X.216 Presentation service definition for open systems interconnection for CCITT applications

X.217 Association control service definition for open systems interconnection for CCITT applications

X.218 Reliable transfer: model and service definition

X.219 Remote operations: model, notations and services definition

X.220 Use of X.200-series protocols in CCITT applications

X.223 Use of X.25 to provide the OSI connection-mode network service for CCITT applications

X.224 Transport protocol specification for open systems interconnection

X.225 Session protocol specification for open systems interconnection for CCITT applications

X.226 Presentation protocol specification for open systems interconnection

X.227 Association control protocol specification for open systems interconnection for CCITT applications

X.228 Reliable transfer: protocol specification

X.229 Remote operations: protocol specification

X.244 Procedure for the exchange of protocol identification during virtual call establishment on packet switched PDNs

X.290 OSI conformance testing methodology and framework for protocol recommendations for CCITT applications

X.300   General principles for interworking between public networks, and between public networks and other networks for the provision of data transmission services

X.301   Description of the general arrangements for the call control within a subnetwork and between subnetworks for the provision of data transmission services

X.302   Description of the general arrangements for internetwork utilities within a subnetwork and intermediate utilities between subnetworks for the provision of data transmission services

X.305   Functionalities of subnetworks relating to the support of the OSI connection-mode network service

X.320   General arrangements for interworking between ISDNs for the provision of data transmission services

X.321   (I.540) General arrangements for interworking between CSPDNs and ISDNs for the provision of data transmission services

X.322   General arrangements for interworking between PSPDNs and CSPDNs for the provision of data transmission services

X.323   General arrangements for interworking between PSPDNs

X.324   General arrangements for interworking between PSPDNs and public mobile systems for the provision of data transmission services

X.325   (I.550) General arrangements for interworking between PSPDNs and ISDNs for the provision of data transmission services

X.326   General arrangements for interworking between PSPDNs and the common channel signalling network (CCSN)

X.327   General arrangements for interworking between PSPDNs and private data networks for the provision of data transmission services

X.350   General interworking requirements to be met for data transmission in international public mobile satellite systems

X.351   Special requirements to met for packet assembly/disassembly facilities (PADs) located at or in association with coast earth stations in the public mobile satellite service

X.352   Interworking between PSPDNs and public maritime mobile satellite data transmission service

X.353   Routing principles for interconnecting public maritime mobile satellite data transmission systems with PDNs

X.370   Arrangements for the transfer of internetwork management information

X.400   Message handling systems: system model-service elements

X.401   Message handling systems basic service elements and optional user facilities

X.402   Message handling systems: overall architecture

X.403   Message handling systems: conformance testing

X.407   Message handling systems: abstract service definition conventions

X.408   Message handling systems: encoded information type conversion rules

X.409   Message handling systems: presentation transfer syntax and notation

X.410   Message handling systems: remote operation and reliable transport server

X.411   Message handling systems: message transport layer

X.413   Message handling systems: message store: abstract service definition

X.419   Message handling systems: protocol specifications
X.420   Message handling systems: interpersonal messaging user agent layer
X.430   Message handling systems: access protocol for teletex terminals
X.435   Message handling systems: EDI messaging system
X.500   The directory: overview of concepts, models and services
X.501   The directory: models
X.509   The directory: authentication framework
X.511   The directory: abstract service definition
X.518   The directory: procedures for distributed operation
X.519   The directory: protocol specifications
X.520   The directory: selected attribute types
X.521   The directory: selected object classes

Appendix 4. ISO Standards

Because of the rapid development of standardization activities, this list does not distinguish finally approved standards from their drafts and, possibly, draft proposals.

In the interests of brevity, the prefix IS is dropped and the following abbreviations are used in this appendix:

Add   addendum
DC   data communication
DP   data processing
IP   information processing
IPS   information processing systems
IT   information technology
LAN   local area networks
OSI   open systems interconnection
Pt   part
TC   text communication
TR   technical report

646	IPS	ISO seven-bit coded character set for information interchange
907	TR	Local area networks CSMA/CD 10 Mbit/s baseband planning and installation guide
963	IP	Guide for the definition of four-bit character sets derived from the seven-bit coded character set for information processing interchange
1155	IP	Use of longitudinal parity to detect errors in information messages
1177	IP	Character structure for start–stop and synchronous character oriented transmission
1745	IP	Basic mode control procedures for data communication systems
2022	IP	ISO seven-bit and eight-bit coded character sets. Code extension techniques
2047	IP	Graphical representation for the control characters of the seven-bit coded character set
2110	DC	25-pole DTE–DCE interface connector and contact number assignments

2111	DC	Basic mode control procedures. Code independent information transfer
2382	DP	Vocabulary
		Pt 09: DP Data communication
		Pt 18: IPS Distributed data processing
		Pt 25: IPS Local area networks
		Pt 26: IT OSI architecture
2593	DC	34-pin DTE–DCE interface connector and pin assignments
2628		Basic mode control procedures. Complements 2629
2629		Basic mode control procedures. Conversational information message transfer
3309	IPS-DC	High level data link control procedures. Frame structure
4335	IPS-DC	High level data link control procedures. Consolidation of elements of procedures
		Add 2: Flow-controlled unnumbered information (FUI) frame
		Add 3: Start–stop transmission
4873	IP	ISO eight-bit code for information interchange. Structure and rules for implementation
4902	DC	37-pin and 9-pin DTE–DCE interface connectors and pin assignments
4903	DC	15 pin DTE–DCE interface connectors and pin assignments
6093		Representation of numerical values in character strings
6429	IP	Control functions for seven-bit and eight-bit character sets
6936	DP	Conversion between the ISO seven-bit coded character set (ISO 646) and the CCITT International Telegraph Alphabet No.2 (ITA 2)
6937	IP	Coded character sets for text communication
		Pt 1: General introduction
		Pt 2: Latin alphabetic and non-alphabetic graphic characters
		Pt 3: Control functions for page-image format
		Pt 4: Control functions for formatted and formattable text
		Pt 5: Scientific and technical graphic characters
		Pt 6: Publishing and box drawing graphic characters
		Pt 7: Greek graphic characters
		Pt 8: Cyrillic graphic characters
7064	DC	Check character systems
7350	TC	Registration of graphic character subrepertoires
7477	TR-DC	Arrangement for DTE to DTE physical connection using V.24 and X.24 interchange circuits
7478	IPS-DC	Multilink procedures
7480	IP	Start–stop transmission signal quality at DTE–DCE interface

7498	IPS-OSI	Basic reference model
		Add 1: Connectionless-mode data transmission
		Add 2: Multipeer data transmission
		Pt 2: Security architecture
		Pt 3: Naming and addressing
		Pt 4: Management framework
7776	IPS-DC	High level data link control procedures. Description of the X.25 LAPB-compatible DTE data link procedures
7809	IPS-DC	High level data link control procedures. Consolidation of classes of procedures
		Add 2: Description of optional functions
		Add 3: Start–stop transmission
		Add 4: List of standard data link layer protocols that utilize HDLC classes of procedures
7826		Data interchange: general structure of the interchange of coded representation
7942	IPS	Computer graphic. Functional description
8072	IPS-OSI	Transport service definition
		Add 1: Connectionless-mode transmission
8073	IPS-OSI	Connection oriented transport protocol specification
		Add 1: Network connection management sub-protocol
		Add 2: Class four operation over connectionless network service
		Add 3: Protocol implementation conformance statement performance
8208	IPS-DC	X.25 packet level protocol for data terminal equipment
		Add 1: Alternative logical channel number allocation
		Add 2: Extension for private and switched use
		Add 3: Conformance requirements
8211		Data descriptive file for information interchange
8326	IPS-OSI	Session service definition
		Add 1: Symmetric synchronization for the session protocol
		Add 2: Incorporation of unlimited user data
		Add 3: Connectionless-mode session service
8327	IPS-OSI	Session protocol specification
		Add 1: Symmetric synchronization for the session protocol
		Add 2: Incorporation unlimited user data
8348	IPS-DC	Network service definition
		Add 1: Connectionless-mode transmission
		Add 2: Network layer addressing
		Add 3: Additional features of the network service
8372	IP	Modes of operation for 64-bit block cypher algorithm
8471	IPS-DC	High level data link control balanced class of procedures. Data link layer address resolution/negotiation in switched environment
8473	IPS-DC	Protocol for providing the connectionless-mode network service

		Add 1: Provision of the underlying service assumed by ISO 8473 over point-to-point subnetworks which provide the OSI data link service

Add 1: Provision of the underlying service assumed by ISO 8473 over point-to-point subnetworks which provide the OSI data link service

Add 2: Formal description of ISO 8473

Add 3: Provision of the underlying service assumed by ISO 8473 over subnetworks which provide the OSI data link service

8480	DC	DTE–DCE interface back-up control operation using the 25-pin connector
8481		DTE to DTE physical connection using X.24 interchange circuits with DTE provided timing
8482	IPS-DC	Twisted pair multipoint interconnections
8505	IP-TC	Functional description and service specification for message oriented text interchange systems
8509	TR-IPS	OSI service convention
8571		File transfer, access and management

Pt 1: General description

Pt 2: Virtual filestore

Pt 3: File service definition

Pt 4: File protocol specification

Pt 5: Protocol implementation conformance statement performance

8602	IPS-OSI	Protocol for providing the connectionless-mode transport service
8632	IPS	Computer graphics. Metafile for the storage and transfer of picture description information

Pt 1: Functional specification

Pt 2: Virtual filestore

Pt 3: Binary encoding

Pt 4: Cleartext coding

8648	IPS-OSI	Internal organization of the network layer
8649	IPS-OSI	Service definition for the association control service element

Add 1: Peer-entity authentication during association establishment

Add 2: Connectionless-mode ACSE service

8650	IPS-OSI	Protocol specification of the association control service element

Add 1: Peer-entity authentication during association establishment

8651	IPS	Computer graphics. Graphical kernel system language binding

Pt 1: FORTRAN

Pt 2: PASCAL

8802	IPS-LAN	Local area networks

Pt 1: General introduction

Pt 2: Logical link control

Add 1: Flow control technique for multi-segment networks

		Add 2: Acknowledged connectionless-mode service, type 3 operation
		Pt 3: Carrier sense multiple access with multiple collision
		Add 1: Medium attachment unit and baseband medium specification for type 10BASE2
		Add 2: Repeater unit specification for use with 10BASE5 and 10BASE2 networks
		Add 3: Broadband medium attachment unit and broadband medium specification type 10BROAD36
		Pt 4: Token-passing bus access method and physical layer specification
		Pt 5: Token ring access method and physical layer specification
		Pt 7: Slotted ring access method and physical layer specification
8806		Computer graphics. Graphical kernel system for three-dimension functional description
		Pt 1: FORTRAN
8807	IPS-OSI	LOTOS A formal description technique based on the temporal ordering of observational behaviour
8822	IPS-OSI	Connection oriented presentation service definition
		Add 1: Connectionless-mode presentation service
8823	IPS-OSI	Connection oriented presentation protocol specification
		Add 1: Protocol implementation conformance statement
8824	IPS-OSI	Specification of abstract syntax notation one (ASN.1)
8825	IPS-OSI	Specification of basic encoding rules for abstract syntax
		Add 1: ASN.1 extension
8831	IPS-OSI	Job transfer and manipulation (JTM): concepts and services
		Pt 2: JTM concepts and services
8832	IPS-OSI	Job transfer and manipulation (JTM): basic class protocol specification
		Add 1: Extension to specification of the full protocol
8859	IP	Eight-bit single byte coded graphic character sets
		Pt 1: Latin alphabet No.1
		Pt 2: Latin alphabet No.2
		Pt 3: Latin alphabet No.3
		Pt 4: Latin alphabet No.4
		Pt 5: Latin/Cyrillic alphabet
		Pt 6: Latin/Arabic alphabet
		Pt 7: Latin/Greek alphabet
		Pt 8: Latin/Hebrew alphabet
		Pt 9: Latin alphabet No.5
8867		Industrial asynchronous data link and physical layer
		Pt 1: Physical interconnection and two-way alternate communication

8877	IPS	Interface connector and contact assignment for ISDN basic access interface located at reference points S and T
		Add 1: Standard ISDN basic access TE connecting cord
8878	IPS-DC	Use of X.25 to provide the OSI connection-mode network service
		Add 1: Protection and priority
		Add 2: Use of an X.25 PVC to provide the OSI connection-mode
8880	IPS	Protocol combinations to provide and support the OSI network service
		Pt 1: General principles
		Pt 2: Provision and support of the connection-mode network service
		Pt 3: Provision and support of the connectionless-mode network service
8881	IPS-DC	Use of the X.25 level protocol in local area networks
		Pt 1: Use with LLC Type 1 procedures
		Pt 2: Use with LLC Type 2 procedures
8882	IPS	X.25 DTE conformance testing
		Pt 1: General principles
		Pt 2: Data link layer test suite
		Pt 3: Packet layer conformance test
8883	IP-TC	Message oriented text interchange system, message transfer sublayer, message interchange service and message transfer protocol
8885	IPS-DC	High-level data link control procedures. General purpose XID frame information field content and format
		Add 1: Additional operational parameters for the parameter negotiation data link layer subfield and definition
		Add 2: Start–stop transmission
		Add 3: Definition of a private parameter data link layer subfield
		Add 4: Extended transparency options for start–stop transmission
		Add 7: Frame check sequence negotiation
8886	IPS-DC	Data link service definition for open systems interconnection
9036	IP	Arabic seven-bit coded character set for information interchange
9040	IPS-OSI	Virtual terminal basic class service
		Add 1: Extended facility service
9041	IPS-OSI	Virtual terminal basic class protocol
		Pt 1: Specification
9065	IP-TC	Message oriented text interchange system user agent sublayer: interpersonal messaging user agent: message interchange formats and protocols

9066	IPS-TC	Reliable transfer
		Pt 1: Model and service definition
		Pt 2: Protocol specification
9067	IPS-DC	Automatic fault isolation procedures using test loops
9068	IPS	Provision of the connectionless network service using ISO 8208
9072	IPS-TC	Remote operation
		Pt 1: Model, notation and service definition
		Pt 2: Protocol specification
9074	ESTELLE	A formal description technique based on an extended state transmission model
9160	IP	Data encipherment: physical layer interoperability requirements
9307	IP	Data encipherment: specification of DEA 2, a public key algorithm
9314	IPS	Fibre distributed data interface (FDDI)
		Pt 1: Physical layer protocol
		Pt 2: Media access control
		Pt 3: Physical layer medium dependent
9322		Intelligent peripheral interface: generic command set for communication
9496	IP	Programming languages: CCITT high level language (CHILL)
9542	IPS-DC	End system to intermediate system routing exchange protocol for use in conjunction with the protocol for the provision of the connectionless-mode network service (ISO 8473)
9543	IPS	Information exchange between systems: synchronous transmission signal quality at DTE–DCE interfaces
9545	IPS-OSI	Application layer structure
9548	IPS-OSI	Session connectionless protocol to provide connectionless-mode session service
9549	IPS	Galvanic isolation of balanced interchange circuits
9571	TR-IPS-OSI	LOTOS description of the session service
9572	TR-IPS-OSI	LOTOS description of the session protocol
9574	IPS-DC	Provision of the OSI connection-mode network service by packet mode terminal equipment connected to an ISDN
9575	TR-IT	Telecommunications and information exchange between systems: OSI routing framework
9576	IPS-OSI	Connectionless presentation protocol specification
9577	TR	Protocol identification in the network layer
9578	TR	Communication interface connectors used in LAN
9592	IPS	Computer graphics: programmer's hierarchical interactive graphic system
		Pt 1: Functional description
		Pt 2: Archive file format
		Pt 3: Cleartext encoding of archive file
9594	IPS-OSI	The directory
		Pt 1: Overview of concepts, models and service

		Pt 2: Models
		Pt 3: Abstract service definition
		Pt 4: Procedures for distributed operations
		Pt 5: Protocol specifications
		Pt 6: Selected attribute types
		Pt 7: Selected object classes
		Pt 8: Authentication framework
9595	IT-OSI	Common management information service definition
		Pt 2: Common management information service definition
9596	IT-OSI	Common management information protocol specification
		Pt 2: Common management information protocol specification
9636	IPS	Computer graphics: interface technique for dialogues with graphical devices. Computer graphic interface
		Pt 1: Overview
		Pt 2: Control, negotiation and errors
		Pt 3: Output and attributes
		Pt 4: Segments
		Pt 5: Input and echoing
		Pt 6: Raster
9646	IT-OSI	Conformance testing methodology and framework
		Pt 1: General concepts
		Pt 2: Abstract test suite specification
		Pt 3: The tree and tabular combined notation
		Pt 4: Test realization
		Pt 5: Requirements on test laboratories and clients for the conformance assessment process
9798	IPS	Data cryptographic techniques: peer entity authentication mechanism using an n-bit secret key algorithm
		Pt 1: General model for peer entity authentication mechanism
9799		Peer entity authentication using a public key algorithm with a two-way handshake
9804	IPS-OSI	Service definition for the commitment, concurrency and recovery service element
9805	IPS-OSI	Protocol specification for the commitment, concurrency and recovery service element
9834	IPS-OSI	Procedures for the operation of OSI registration authorities
		Pt 1: General procedures
		Pt 3: Registration of object identifiers component values for joint ISO-CCITT use
		Pt 4: Register of virtual terminal profiles
		Pt 5: Register of virtual terminal control object definitions
		Pt 6: Registration of AP titles and AE titles
10016	IP	Modes for operation for an n-bit block cipher algorithm

10021	IPS-TC	Message oriented text interchange system
		Pt 1: System and service overview
		Pt 2: Overall architecture
		Pt 3: Abstract service definition connection
		Pt 4: Message transfer system: abstract service defin- itions and procedures
		Pt 5: Message store: abstract service definition
		Pt 6: Protocol specification
		Pt 7: Interpersonal messaging user agent layer
10022	IPS-OSI	Physical service definition
10023		Formal description of ISO 8073 in LOTOS
10024		Formal description of transport of ISO 8073 specifica- tion in LOTOS
10025		Procedures for testing conformance to ISO 8073
		Pt 1: General principles
10026	IPS-OSI	Transaction processing
		Pt 3: Protocol specification
10028		OSI network layer: intermediate systems functions
10029	IT	Telecommunications and information exchange be- tween systems: operation of an X.25 interworking unit
10030	IPS	End system to intermediate system routing exchange protocol for use in conjunction with ISO 8208 (X.25/ PLP)
10031	IP-TC	Distributed office applications model
		Pt 1: General model
		Pt 2: Referenced data transfer
10032	IPS	Reference model of data management
10035	IPS-OSI	Connectionless associated control service element protocol specification
10038	IPS-LAN	MAC sublayer interconnection (MAC bridging)
10039	IPS-LAN	MAC service definition
10040	IT-OSI	System management overview
10117		Peer entity authentication mechanism using a public key algorithm with a three-way handshake
10148	IPS	Basic remote procedure call using OSI remote operation
10164	IT-OSI	Systems management
		Pt 1: Object management function
		Pt 2: State management function
		Pt 3: Attributes for representing relationships
		Pt 4: Alarm reporting function
		Pt 5: Event report management function
		Pt 6: Log control function
		Pt 7: Security alarm reporting function
10165	IT-OSI	Management information services, structure of man- agement information
		Pt 1: Management information model
		Pt 2: Definition of management information
		Pt 4: Guidelines for the definition of managed objects

10168	IT-OSI	Conformance test suite for the session protocol Pt 1: Test suite structure and test purposes
10169	IT-OSI	Conformance test suite for the ACSE protocol Pt 4: Test suite structure and test purposes
10171	TR	List of standard data link layer protocols that utilize high-level data link control (HDLC) classes of procedures
10173	IT	ISDN primary access connector at reference points S and T
10177	IPS-DC	Intermediate system support of the OSI connection-mode network service using ISO 8208 in accordance with ISO 10028
10181	IPS-OSI	Security frameworks for open systems
10367		Repertoire of standardized coded graphic character sets for use in eight-bit codes

Appendix 5. Abbreviations

Some items are accompanied by references (in parentheses) where the abbreviation is mostly used and, if needed, by another abbreviation with the same meaning. The note "prefix" means that the abbreviation is used as a prefix of another abbreviation.

A	adaptor, additional user facility, availability, application (prefix)
A, AD, ADR	address
AAS	abbreviated address signal
ABM	asynchronous balance mode (HDLC)
AC	accept
ACAU	automatic calling and answering unit
ACE	automatic calling equipment (ACU, ADU)
ACF	advanced communication function
ACK	acknowledge (IA5)
ACR	abandon call and retry (circuit)
ACS	advanced communication services
ACSE	association control service element
ACU	automatic calling unit (ACE, ADU)
ADCCP	advanced data communication control procedure
ADM	asynchronous disconnect mode (HDLC)
ADMD	administration management domain
ADTE	asynchronous DTE
ADU	automatic dialling unit (ACE, ACU)
AFI	authority and format identifier
AI	activity interrupt
AK	data acknowledgement
AM	access module
ANSC	American National Standards Committee
ANSI	American National Standards Institute
AOW	Asia Oceania Workshop
AP	application process
APA	active position addressing
API	asynchronous pollable interface
APR	alternate path retry
AR	activity resume
ARM	asynchronous response mode (HDLC)
ARQ	automatic request for repetition, automatic repeat request

ARU	automatic response unit
AS	address signal, activity start
ASC	Accredited Standards Committee
ASCII	American standard code for information interchange
ASN.1	abstract syntax notation one
ATDM	asynchronous time division multiplexer
ATM	advanced testing methods, asynchronous transfer mode (ISDN)
AT&T	American Telephone and Telegraph Company
AU	answer unit; access unit (ISDN)
B	byte; byte timing (circuit)
BAS	basic activity subset
BC	binary checksum
BCC	block check character
BCD	binary coded decimal
BCK	block
BCR	block coding rate
BCS	block check sequence, basic combined subset
BCUG	bilateral closed user group
BCUGO	bilateral closed user group with outgoing access
BEL	bell (IA5)
BER	bit error rate
BIC	bearer identification code
B-ISDN	broadband ISDN
BISYNC	binary synchronous communication (BSC)
BOP	bit-oriented protocol
BP	block parity
BPS	bits per second
BR	bit rate
BS	backspace (IA5)
BSC	binary synchronous communication (BISYNC)
BSI	binary synchronous interface
BSS	basic synchronized subset
BT	business telecommunications, British Telecom
BTAM	basic telecommunication access method
BW	both-way
C	call, connect (CN)
CA	communication adapter
CAC	call accepted (packet)
CAD	call acceptance delay
CAM	communication access method, call accepted message
CAN	cancel (IA5)
CAR	call request (packet)
CASE	common application service element
CC	call connected packet, communication computer, country code, congestion control, connection confirm
CCB	cyclic check byte
CCC	communication control character
CCIF	International Telephone Consultative Committee (Comité Consultatif International Téléphonique)

CCIR	International Radio Consultative Committee (Comité Consultatif International des Radiocommunications)
CCIT	International Telegraph Consultative Committee (Comité Consultatif International Télégraphique)
CCITT	International Telegraph and Telephone Consultative Committee (Comité Consultatif International Télégraphique et Téléphonique)
CCP	character controlled protocol
CCS	common channel signalling
CCSN	common channel signalling network
CCU	communication control unit
CDI	called line identity
CDT	connectionless data transmission
CE	customer equipment (CPE)
CEN	European Committee for Standardization (Comité Européen de Normalisation)
CENELEC	European Committee for Electrotechnical Standardization (Comité Européen de Normalisation Electrotechnique)
CEPT	European Conference of Postal and Telecommunications Administrations (Conference Européenne des Administrations des Postes et des Télécommunications)
CIA	channel interface adapter
CIC	circuit identification code
CK	check bit
CL	connectionless (CNL, DG)
CLC	clear confirmation (packet)
CLI	clear indication (packet)
CLR	clear request (packet)
CMB	cyclic redundancy check message block
CMC	communication mode control
CMDR	command reject (HDLC)
CN	connect (HDLC) (C)
CNIC	clearing network identification code
CNL	connectionless (CL, DG)
CNM	communication network management
CO	connection
COAM	customer owned and maintained
CON	call connected (packet); connect message; connection oriented
CONS	connection-mode network service
CP	communication processor
CPE	customer provided equipment (CE)
CPG	call progress message (ISDN)
CPU	central processing unit
CR	connection request packet; call request; carriage return (IA5)
CRC	cyclic redundancy check
CRCC	cyclic redundancy check character
CRP	call request packet
CS	circuit switched
CSC	control signalling code

CSCC	command session change control
CSI	called subscriber identification
CSMA	carrier sense multiple access
CSMA/CD	carrier sense multiple access with collision detection
CSPDN	circuit switched public data network
CT	circuit
CTD	cumulative transit delay
CTT	code transparent transmission
CU	control unit
CUG	closed user group
CWP	communicating word processor
D	delivery confirmation bit, disconnect
DA	digital access
DAA	data access arrangement
DC	data concentrator, device control, disconnect confirm (prefix)
DCA	distributed communications architecture (UNIVAC)
DCC	data country code
DCE	data circuit terminating equipment
DCS	digital command signal
DDD	direct distance dialling
DCU	data control unit
DDI	direct dialling in
DIF	document interchange format
DEA	data encryption algorithm
DEC	Digital Equipment Corporation
DEE	data encryption equipment
DEMUX	demultiplexer
DES	data encryption standard
DFC	data flow control
DG	datagram (CL, CNL)
DIS	draft international standard (ISO)
DISC	disconnect (HDLC)
DL	data link (prefix)
DLC	data link control
DLCC	data link control chip
DLE	data link escape (IA5)
DLO	data line occupied
DM	disconnect mode (HDLC)
DN	disconnect
DNA	digital network architecture (DEC)
DNIC	data network identification code
DP	draft proposal (ISO)
DR	disconnect request, discount request (prefix)
DS	data set, document storage
DSE	data switching exchange
DSP	display station protocol, domain specific port
DSS	digital subscriber signalling
DST	destination reference (field)
DSU	data switching unit
DT, DTA	data

DTAM	document transfer and manipulation
DTE	data terminal equipment
DU	data unit
DUP	data user part
DXS	data exchange system
E	essential user facility, expedited (prefix) (X)
EA	expedited data acknowledgement
EBCDIC	extended binary coded decimal interchange code
EBU	European Broadcasting Union
EC	echo check
ECC	error control code, error correction code
ECMA	European Computer Manufacturers Association
ECSA	Exchange Carriers Standards Association
ED	error detection, expedited data (EX)
EDC	error detection and correction (ERCC)
EDAC	error detection and automatic correction
EDI	electronic data interchange
EDS	electronic data switching system
EFdS, EFDS	error-free deciseconds
EFS	error-free seconds
EFT	electronic funds transfer
EIA	Electronic Industries Association (in the USA)
EM	end of medium (IA 5)
ENQ	enquiry (IA5)
EOT	end of transmission (IA5)
EPOS	electronic point of sale
ER	error, exception report
ERC	error control
ERCC	error checking and correction (EDC)
ERP	error recovery procedure
ERR	error, protocol data unit error
ESC	escape (IA5)
ESS	electronic switching system
ESTELLE	extended state transition language
ET	exchange termination (ISDN)
ETB	end of transmission block (IA5)
ETS	European Telecommunications Standard
ETSI	European Telecommunications Standards Institute
ETX	end of text (IA5)
EU	end user
EVP	envelope
EWOS	European Workshop on Open Systems
EX	expedited data (ED)
F	final bit (HDLC), flag
FAX	facsimile
FCC	Federal Communications Commission (in the USA), flow control confirmation
FCS	frame check sequence (HDLC)
FDDI	fibre distributed data interface
FDM	frequency division multiplex

FDMA	frequency division multiple access
FDT	formal description technique
FDX	full duplex
FE	format effector (IA5)
FEC	forward error control, forward error correction
FEP	front-end processor
FF	form feed (IA5)
FIPS	federal information processing standard (in the USA)
FN	finish
FR	fragment, frame
FRMR	frame reject (HDLC)
FS	file separator (IA5)
FTAM	file transfer, access and management
G	ground, group
GA	go ahead
GAS	Special Autonomous Group (Group Autonome Spécialisé)
GFI	general format identifier
GOS	grade of service
GS	group separator (IA5)
GSM	special mobile group (Groupe Spécial Mobil)
GSTN	general switched telephone network
GT -	give tokens (prefix)
GW	gateway
HDLC	high level data link control procedure
HDR	header
HDX	half duplex
HF	human factor
HLF	high layer function
HOST	host computer
HT	horizontal tabulation (IA5)
I	interface, information field (HDLC), indication (circuit), information
IA	international alphabet, incoming access
IC	incoming call
ICB	incoming calls barred
ICD	international code designator
ICP	initial connection protocol
ID	identification, identifier
IDA	integrated digital access
IDI	initial domain identifier
IDN	integrated digital network
IDP	initial domain part internet datagram protocol
IDSE	international DSE, interworking data switching exchange
IEC	International Electrotechnical Commission
IEEE	Institute of Electrical and Electronic Engineers (in the USA)
I-ETS	Interim European Telecommunications Standard
IFIP	International Federation of Information Processing
IFRB	International Frequency Registration Board
IM	initialization mode (HDLC)
IMP	interface message processor

INC	incoming call (packet)
INIC	ISDN network identification code
INMC	international network management centre
INT	interrupt (packet)
INTC	interrupt confirmation (packet)
I/O	input/output
IP	interface processor, ISDN port, interpersonal (prefix)
IPDN	international public data network
IPMS	interpersonal messaging system
IPRC	Intellectual Property Rights Committee
IPSS	international packet switching service
IPT	interactive protocol tester
IS	information separator (IA5), international standard (ISO)
ISDN	integrated services digital network
ISM ISDN	ISDN Standards Management Special Committee
ISO	International Organisation for Standardisation
ISP	international standard profile
ISUP	ISDN user part
ITA	international telegraph alphabet
ITU	International Telecommunication Union
IU	interface unit
IWF	interworking function
IWU	interworking unit
JTC	Joint Technical Committee
JTM	job transfer and manipulation
LAN	local area network
LAP	link access procedure (HDLC)
LAPB	link access procedure balanced (HDLC)
LAPD	LAP on the D-channel (ISDN)
LC	line concentrator, link control
LCM	limited conversation mode
LCN	logical channel number
LCU	line control unit
LF	line feed (IA5)
LFC	local functional capability
LI	length indicator
LIFO	last in–first out
LIU	line interface unit
LL	link level
LLC	logical link control
LNI	local network interface
LOTOS	language for temporal ordering specification
LPU	line processing unit
LRC	longitudinal redundancy check
LSB	least significant bit
LWE	lower window edge
M	mandatory, modifier function bit (HDLC), more data, modem, multilink (prefix)
MAC	media access control
MADS	multiple access data system

MAF	multiple access facility
MAN	metropolitan area network
MATD	maximum acceptable transit delay
MD	management domain
M/D	modulator/demodulator
MDNS	managed data network services
MH	message handling
MHS	message handling service, message handling system
MIB	management information base
MIPS	management information protocol specification
MISD	management information service definition
MLC	multilink control field
MLI	multi-leaving interface
MLP	multilink procedure
MML	man–machine language
MPDU	message protocol data unit
MRT	mean repair time
MSB	most significant bit
MSDSE	maritime satellite data switching exchange
MSS	maritime satellite service, mobile satellite system
MT	message transfer
MTA	message transfer agent
MTBCF	mean time between component failures
MTBF	mean time between failures
MTBSO	mean time between service outages
MTP	message transfer part
MTS	message transfer system
MTTR	mean time to repair
MTTSR	mean time to service restoral
muldex	multiplexer/demultiplexer
MUX	multiplexer (MX)
MX	multiplexer (MUX)
N	network (prefix), normal (prefix)
NA	network adaptor (ISDN), network address, network aspects
NAD	network access device
NAE	network address extension
NAK	negative acknowledge (IA5)
NAS	network administration system
NBS	National Bureau of Standards (in the USA, now NIST)
NC	network congestion, network connection
NCC	network control centre
NCCD	network-dependent call connection delay
NCP	network control program
NCU	node control unit
NDC	national destination code
NDM	normal disconnect mode (HDLC)
NDN	non-delivery notification
NET	European Telecommunications Standard (Norme Européenne de Télécommunications)
NI	network identifier

NIL	network interface language
NIST	National Institute for Science and Technology (in the USA, formerly NBS)
NL	network layer, new line (IA5)
NMC	network management centre
NMCS	network management and control system
NN	national number
NNI	network node interface (ISDN)
NP	network performance
NPAI	network protocol address information
N(R)	transmitter receive sequence number (HDLC)
NRC	network routing centre
NRM	normal response mode (HDLC)
NS	network service
N(S)	transmitter send sequence number (HDLC)
NSAP	network service access point
NSC	network switching centre
NSF	non-standard facility
NSP	network service part, network service protocol
NT	network termination (ISDN)
NTN	network terminal number
NTS	network test system
NUI	network user identification
NUL	null (IA5)
NVT	network virtual terminal
NW	network
OA	outgoing access
OAM, OA&M	operation, administration and maintenance
OCB	outgoing calls barred
OMC	operation and maintenance centre
ONSD	optional network specific digit
OPI	originating point code
O/R	originator/recipient
OSA	open systems architecture
OSI	open systems interconnection
OSIE	OSI environment
OSI/RM	OSI reference model
OWC	one-way communication
P	poll bit (HDLC), presentation (prefix), protocol
PABX	private automatic branch exchange
PAD	packet assembler and disassembler
PAI	protocol addressing information
PAR	parameter, positive acknowledgement and retransmission
PB	parity check block
PBX	private branch exchange
PCI	protocol control information
PCM	pulse code modulation
PDN	packet data network, public data network
PDTE	packet DTE
PDU	protocol data unit

PE	packet layer reference event
PH	packet handler, packet handling
Ph	physical (prefix)
PI	parameter identifier
PKE	public key encryption
PL	private line
PLACO	Planning Committee (in ISO)
PLMN	public land mobile network
PLP	packet layer protocol
PNIC	private data network identification code
PO	post office (PTT)
POS	point of sale
P(R)	packet receive sequence number (HDLC)
PRE	prefix
PRMD	private management domain
PS	packet switched, packet switching
P(S)	packet send sequence number (HDLC)
PSDN	public switched data network, packet switched data network
PSDTS	packet switched data transmission service
PSE	packet switching exchange
PSN	packet switched network, public switched network
PSPDN	packet switched public data network
PSTN	public switched telephone network
PT	parallel transmission, project team (ETSI)
PTI	packet type identifier
PTT	postal, telegraph and telephone (PO)
PV	parameter value
PVC	permanent virtual circuit
PvtDN	private data network
Q	qualifier bit (HDLC)
QOS	quality of service
QRP	QOS reference point
QTAM	queued telecommunication access method
R	recommendation, receive (circuit), release, response
RA	resynchronize acknowledgement
RD	request for disconnect (HDLC)
REC	recovery, request confirmation (packet)
REF	reference
REI	reset indication (packet)
REJ	reject (HDLC, packet)
REL	release (ISDN)
RER	residual error rate
RES	reset request (packet)
RESET	resetting
RIM	request for initialization mode (HDLC)
RIT	rate of information throughput
RJ	reject (prefix)
RJE	remote job entry
RNR	receive not ready (HDLC, packet)
ROSE	remote operation service element

RPC	remote procedure call
RPOA	recognized private operating agency
RPT	repeat
RR	receive ready (HDLC, packet)
RS	record separator (IA5), resynchronize
RSA	Rivest, Shamir and Adleman (cryptosystem)
RSET	reset
RSP	response
RTC	restart confirmation (packet)
RTI	restart indication (packet)
RTR	restart request (packet)
RTSE	reliable transfer service element
S	segment, session (prefix), supervisory function bit (HDLC), set-up (ISDN), ISDN access interface
SA	set-up acknowledge (ISDN)
SABM	set asynchronous balanced mode (HDLC)
SABME	set asynchronous balanced mode extended (HDLC)
SAP	service access point
SAPI	service access point identifier
SARM	set asynchronous response mode (HDLC)
SARME	set asynchronous response mode extended (HDLC)
SC	session control, subcommittee (ISO)
SCA	synchronous communication adaptor
SCCP	signalling connection control part
SCPC	single channel per carrier
SDE	submission and delivery entity
SDL	specification and description language, systems definition language
SDLC	synchronous data link control procedure (IBM)
SDTE	synchronous DTE
SDU	service data unit
SFD	simple formattable document
SG	study group (CCITT)
SGR	select graphic rendition
SHS	select horizontal spacing
SI	shift in (IA5), SPDU identifier
SIF	signal information field
SIM	set initialization mode (HDLC)
SLAP	subscriber line access protocol
SLP	single link procedure
SMAE	system management application entity
SMAP	system management application process
SMIB	security management information base
SMVT	scroll mode virtual terminal
SN	subscriber number
SNA	systems network architecture (IBM)
SNACP	subnetwork access protocol
SNDCP	subnetwork dependent convergence protocol
SNICP	subnetwork independent convergence protocol
SNPA	subnetwork point of attachment

SNR	signal to noise ratio
SNRM	set normal response mode (HDLC)
SNRME	SNRM extended (HDLC)
SO	shift out (IA5)
SOH	start of heading (IA5)
SP	space (IA5), signalling point
SPC	stored program control
SPS	signalling protocols and switching
SREJ	selective reject (HDLC)
SS	session service (prefix), single shift
SS No.7	signalling system No.7
SST	start–stop transmission
ST	serial transmission
STC	subtechnical committee (ETSI)
STDM	synchronous time division multiplex
STE	signalling terminal
STM	synchronous transfer mode
STP	signal transfer point
STX	start of text (IA5)
SU	signal unit
SUB	substitute character (IA5)
SVC	switched virtual circuit
SVS	select vertical spacing
SX	simplex
SYN, SYNC	synchronous idle, synchronization (IA5)
T	transmit (circuit), transport (prefix), ISDN access interface
TA	terminal adaptor (ISDN)
TASI	time assignment speech interpolation
TC	technical committee (ETSI, ISO, IFIP), transmission control (IA5), transport connection
TCAP	transaction capabilities application part
TCU	transmission control unit
TDI	transit delay indication
TDM	time division multiplex
TDMA	time division multiple access
TDS	transit delay selection
TE	terminal equipment (ISDN)
TEI	terminal endpoint identifier (ISDN)
TID	terminal identification
TIP	terminal interface processor
TLX	Telex
TM	transparent mode, transmission and multiplexing
TMN	telecommunication management network
TNIC	transit network identification code
TOA	type of address
TOF	transfer overhead factor
TOT	transfer overhead time
TPTD	total data packet network transfer delay
TRAC	Technical Recommendations Application Committee
TRIB	transfer rate of information bits

TRN	transparent (X)
TS	transport service (prefix), transport station, time slot
TTC	Telecommunication Technology Committee (in Japan)
TTD	target transit delay
TTX	teletex
TTXAU	teletex access unit
TTXT	transparent text
TWA	two-way alternate
TWS	two-way simultaneous text utilization, unnumbered (prefix), unavailability
UA	user agent
UCCD	user-dependent call connection delay
UI	unnumbered information (HDLC), unit interval
UM	user message
UNI	user–network interface (ISDN)
UP	unnumbered poll (HDLC), user part
UPTD	user-dependent data packet transfer delay
UWE	upper window edge
VAN	value added network
VAS	value added service
VC	virtual call, virtual circuit
VDT	visual display terminal
VPN	virtual private network
V(R)	receive state variable (HDLC)
VRC	vertical redundancy checking
V(S)	send state variable (HDLC)
VSAM	variable sequential access method
VT	vertical tabulation (IA5), virtual terminal
VTAM	virtual telecommunication access method
VTP	virtual terminal protocol
VTX	videotex
W	window, window size (WS)
WABT	wait before transmission
WAN	wide area network
WG	working group (CCITT, ISO, IFIP)
WP	word processing, working party (CCITT)
WS	window size (W)
X	transparent (TRN), expedited (prefix) (E)
XID	exchange of identification characters (HDLC)
XMAN	extended MAN

References

1. Arango M, Badr H, Gelernter D (1985) Staged circuit switching. IEEE Transactions on Computers C-34:174–180
2. Baran P (1964) On distributed communication networks. IEEE Transactions on Communication Systems CS-12:1–9
3. Barberis G, Guarneri MR, Macrina P (1987) Handling packet services within ISDN. Computer Communications 3:128–133
4. Barnes AC, Cole AS (1986) Network management for public data networks in the UK. In: Kühn P (ed) New communication services: A challenge to computer technology. North-Holland, Amsterdam, pp 457–461
5. Barnet R, Marvard L, Smith S (1987) Packet switched networks: Theory and practice. Wiley, London New York
6. Becker D (1986) Technical aspects of data communication in the ISDN. In: Kühn P (ed) New communication services: A new challenge to computer technology. North-Holland, Amsterdam, pp 17–21
7. Bertsekas D, Gallager R (1987) Data networks. Prentice-Hall, Englewood Cliffs, NJ
8. Bocker P (1988) ISDN: The integrated services digital network (Concepts, methods, systems). Springer, Berlin Heidelberg New York
9. Brunn D (1989) Network management for open systems interconnection through ISDN. In: Kühn PS (ed) Kommunikation in verteilten Systemen. Springer, Berlin Heidelberg New York, pp 703–717
10. Burg FM, Puges P (1988/89) X.25: It's come a long way. Computer Networks and ISDN Systems 16:395–404
11. Clyne L (1989) Lower layer OSI addressing. Computer Networks and ISDN Systems 17:279–281
12. Cohen D, Postel J (1983) The ISO reference model and its protocol architectures. In: Mason REA (ed) Information processing 1983. North-Holland, Amsterdam, pp 29–34 (reprinted 1986)
13. Davies DW (1986) A personal view of the origins of packet switching. In: Csaba L, Tarnay K, Szentivanyi T (eds) Computer network usage: Recent experiences. North-Holland, Amsterdam, pp 1–13
14. Davies DW, Barber DLA, Price WL, Solomonides CM (1980) Computer networks and their protocols. Wiley, Chichester New York Brisbane Toronto
15. Davies DW, Bartlett KA, Scantlebury RA, Wilkinson PT (1967) A digital communication network for computers giving rapid response at remote terminals. In: Proceedings of the ACM symposium on operating system principles. Gatlinburg, October 1967
16. Davies DW, Price WL (1984) Security for computer networks: An introduction to data security in teleprocessing and electronic funds transfer. Wiley, Chichester New York Brisbane Toronto Singapore
17. Deasington RJ (1986) X.25 explained: Protocols for packet switching networks. Wiley, Chichester New York Brisbane Toronto (French translation: Masson, Paris, 1988)
18. Diffie W, Hellman ME (1976) New directions in cryptography. IEEE Transactions on Information Theory IT-22:644–654
19. Ephemerides A (1986) The routing problem in computer networks. In: Blake IF, Poor HV (eds) Communication and networks. Springer, New York, pp 299–324
20. Fak V, Network security and open systems interconnection. In: Khakhar D, Iversen VB (eds) Information network and data communication, vol II. North-Holland, Amsterdam, pp T4/1–T4/10

21. Gallager RG (1981) Applications of information theory for data communication networks. In: Skwirzynski J (ed) New concepts in multi-user communication. NATO Advanced Study Institute, Sijthoff-Noordhoff
22. GAS 11 (1987) GAS 11 Handbook: Strategy for the introduction of a public data network in developing countries. International Telecommunication Union, Geneva
23. Giese E, Görgen K, Hinsch E, Schulze G, Trwöl K (1985) Dienste und Protokolle in Kommunikationssystemen. Springer, Berlin Heidelberg New York London
24. Göran E, Rudberg A, Söderberg L (1989) Intelligent networks. Architectural principles and an economical analysis. In: Lada L (ed) Network Planning in the 1990s. North-Holland, Amsterdam, pp 351–358
25. Grabb DS, Cotton IW (1977) Criteria for evaluation of data communication services. Computer Networks 1:325–340
26. Green PE (1982) Computer network architectures and protocols. Plenum, New York
27. Green PE (1986) Protocol conversion. IEEE Transactions on Communications COM-34:257–268
28. Grimm R (1989) Security on networks: Do we really need it? Computer Networks and ISDN Systems 17:315–323
29. Helling H (1983) Data transmission service tariffs. Telecommunications Journal 50:195–201
30. IEEE Communications Magazine (1990) Special issue on ISDN. Horn RW (ed) 28, April 1990
31. IEEE Journal on Selected Areas in Communications (1989) Architectures and protocols for computer networks: the state of the art. Rudin H, Sabnani K (eds) SAC-7, September 1989
32. IEEE Journal on Selected Areas in Communications (1990) Heterogeneous computer networks interconnection. Green PE, Maemura K, Williamson R (eds) SAC-8, January 1990
33. IEEE Transactions on Communications (1980) Special issue on computer network architectures and protocols. Green PE (ed) COM-28, April, 1980
34. IEEE Transactions on Communications (1981) Special issue on congestion control in computer networks. Rudin H (ed) COM-29, April 1981
35. Irmer T (1989) CCITT's contribution to the evolution of telecommunication networks. In: Lada L. (ed) Network planning in the 1990s. North-Holland, Amsterdam, pp 1–4
36. Israel JE, Weissberger AJ (1987) Communicating between heterogeneous networks. Data Communications 16:215–235
37. Jain BN, Agrawala AK (1990) Open systems interconnection – its architecture and protocols. North-Holland, Amsterdam
38. Jakobs K (1987) OSI addressing strategies. ACM Computer Communications Review 17 3:7–12
39. Kahl P (1989) Tariff pattern for ISDN in the Federal Republic of Germany. In: Arnback J (ed) Innovative services or innovative technology? North-Holland, Amsterdam, pp 387–395
40. Kanbach A, Körber A (1990) ISDN – Die Technik. Huthig, Heidelberg
41. Kbawar A, Unsoy M (1985) Packet switching in ISDN. In: Kühn P (ed) New communication services: A challenge to computer technology. North-Holland, Amsterdam
42. Kessler GC (1990) ISDN: Concepts, facilities and services. McGraw-Hill, Maidenhead
43. Klerer SM (1988) The OSI management architecture. An overview. IEEE Network 2:20–29
44. Kunze H (1986) The ISDN concept of the Deutsche Bundespost and the integration of non-voice services. In: Csaba L, Tarnay K, Szentivanyi T (eds) Computer network usage: Recent experiences. North-Holland, Amsterdam, pp 215–230
45. Macchi C, Guilbert JF (1987) Téléinformatique: Transport et traitement de l'information dans les réseaux et systémes téléinformatiques et télématiques. Dunod, Paris
46. Matsubara MM (1989) Evolution of CCITT numbering plans and network interworking. Computer Networks and ISDN Systems 17:47–57
47. Nussbaumer H (1989, 1990) Computer communication systems: Data circuits, error detection, data links (vol 1); Principles, design, protocols (vol 2). Wiley, Chichester
48. Patel A, Ryan V (1990) Introduction to names, addresses and routes in an OSI environment. Computer Communications 13:27–36
49. Pořízek R (1986) A basis for the taxonomy of error control methods in data communication. Information Systems 15 6:577–590 (in Slovak)
50. Pouzin L (1976) Flow control in data networks – methods and tools. In: ICCC'3. Toronto, August 1976, pp 467–474
51. Proceedings of the IEEE (1978) Special issue on packet switching. Kahn RE, Uncapher KW, VanTrees HL (eds) 66, November 1978
52. Proceedings of the IEEE (1983) Special issue on the OSI. Folts HC, des Jardins R (eds) 71, December 1983

53. Protocols and techniques for data communication networks (1981) Kuo F (ed) Prentice Hall, Englewood Cliffs, NJ
54. Pujolle G, Horlatt E (1990) Architecture des réseaux informatiques: Les outils de communication (vol 1). Eyrolles, Paris
55. Pužman J (1982) Debits and credits in telecommunications. In: Vasko T (ed) Telecommunications: Some policy issues. IIASA, Luxembourg, CP-82–65, pp. 48–60.
56. Pužman J (1986) Network services versus OSI/RM. In: Csaba L, Tarnay K, Szentivanyi T (eds) Computer network usage: Recent experiences. North-Holland, Amsterdam, pp 437–448
57. Pužman J (1987) The architecture model of PDNs: How do they fit in the OSI/RM? In: Proceedings of the first international IBERICOM conference on data communications. Lisboa 1987
58. Pužman J, Pořízek R (1980) Communication control in computer networks. Wiley, Chichester New York Brisbane Toronto
59. Rauch-Hindin W (1985) Communication standards: OSI is not a paper tiger. Systems Software 3:64–86
60. Rivest RL, Shamir A, Adleman L (1971) A method for obtaining digital signature and public key cryptosystems. Communications of the ACM 21:120–126
61. Roberts LG (1967) Multiple computer networks and intercomputer communication. In: Proceedings of the ACM symposium on operating system principles. Gatlinburg, October 1967
62. Rouxeville B (1989) Financial and tariff aspects of ISDN. In: Arnbak J (ed) Innovative services or innovative technology? North-Holland, Amsterdam, pp 383–385
63. Salazar AC, Scarfo PJ, Horn RJ (1987) Network management systems for data communications. IEEE Communications Mag 8:21–27
64. Salomaa A (1990) Public-key cryptography. Springer, Berlin Heidelberg New York London
65. Schwarts M (1987) Telecommunication networks: Protocols, modelling and analysis. Addison-Wesley, Reading, Mass
66. Seitz NB, Bodson D (1986) Data communication performance assessment. Telecommunications 14:33–45
67. Shannon CE (1949) Communication theory of secrecy systems. BSTJ 28:656–715
68. Sluman S (1989) A tutorial on OSI management. Computer Networks and ISDN Systems 17:270–278
69. Stalling W (1985) Integrated services digital network (ISDN). IEEE Press, North-Holland, Amsterdam
70. Stalling W (1988) Data and computer communications. 2nd edn. Macmillan, London
71. Staudinger W (1989) The evolution of a packet switching concept for the ISDN from the view of CCITT, reflected on the activities of the Deutsche Bundespost. In: Arnbak J (ed) Innovative services or innovative technology? North-Holland, Amsterdam, pp 243–254
72. Tanenbaum AS (1983) Computer networks. Prentice-Hall, Englewood Cliffs, NJ
73. Temple S (1991) A revolution in European telecommunications standards making. ETSI, Sophia Antipolis
74. Walford RB (1990) Network system architecture. Addison-Wesley, Reading, Mass
75. Weissberger AJ, Israel JE (1987) What the new working standards provide. Data Communications 16:141–146
76. Weldon EJ (1982) An improved selective repeat ARQ strategy. IEEE Transactions on Communications COM-30:480–486
77. Widl W (1988) Standardization of telecommunication management networks. Ericsson Review 1:17–23
78. Williamson J (1987) Can packet switching survive ISDN? Telephony 213:76
79. Zimmermann H (1981) Development in network architecture and protocols standards. In: Networks. Infotech, Maidenhead, pp 375–392

Subject index